# Science Notebook

Glencoe Science

## Science Level Red

**Consultant**
Douglas Fisher, Ph.D.

New York, New York   Columbus, Ohio   Chicago, Illinois   Peoria, Illinois   Woodland Hills, California

Copyright © Glencoe/McGraw-Hill, a division of The McGraw-Hill Companies, Inc.

P9-APE-619

# About the Consultant

Douglas Fisher, Ph.D., is a Professor in the Department of Teacher Education at San Diego State University. He is the recipient of an International Reading Association Celebrate Literacy Award as well as a Christa McAuliffe award for Excellence in Teacher Education. He has published numerous articles on reading and literacy, differentiated instruction, and curriculum design as well as books, such as *Improving Adolescent Literacy: Strategies at Work* and *Responsive Curriculum Design in Secondary Schools: Meeting the Diverse Needs of Students*. He has taught a variety of courses in SDSU's teacher-credentialing program as well as graduate-level courses on English language development and literacy. He also has taught classes in English, writing, and literacy development to secondary school students.

Copyright © by the McGraw-Hill Companies, Inc. All rights reserved. Permission is granted to reproduce the material contained herein on the condition that such material be reproduced only for classroom use; be provided to students, teachers, and families without charge; and be used solely in conjunction with *Science Level Red*. Any other reproduction, for use or sale, is prohibited without prior written permission of the publisher.

Send all inquiries to:
Glencoe/McGraw-Hill
8787 Orion Place
Columbus, Ohio 43240-4027

ISBN 0-07-874561-6

Printed in the United States of America

11 12 13 14  DOH  15 14 13

Copyright © Glencoe/McGraw-Hill, a division of The McGraw-Hill Companies, Inc.

# Table of Contents

Copyright © Glencoe/McGraw-Hill, a division of The McGraw-Hill Companies, Inc.

# Table of Contents

Copyright © Glencoe/McGraw-Hill, a division of The McGraw-Hill Companies, Inc.

Your notes are a reminder of what you learned in class. Taking good notes can help you succeed in science. These tips will help you take better notes.

- Be an active listener. Listen for important concepts. Pay attention to words, examples, and/or diagrams your teacher emphasizes.

- Write your notes as clearly and concisely as possible. The following symbols and abbreviations may be helpful in your note-taking.

| Word or Phrase | Symbol or Abbreviation | Word or Phrase | Symbol or Abbreviation |
|---|---|---|---|
| for example | e.g. | and | + |
| such as | i.e. | approximately | ≈ |
| with | w/ | therefore | ∴ |
| without | w/o | versus | vs |

- Use a symbol such as a star (★) or an asterisk (*) to emphasis important concepts. Place a question mark (?) next to anything that you do not understand.

- Ask questions and participate in class discussion.

- Draw and label pictures or diagrams to help clarify a concept.

# Note-Taking Don'ts

- **Don't** write every word. Concentrate on the main ideas and concepts.

- **Don't** use someone else's notes—they may not make sense.

- **Don't** doodle. It distracts you from listening actively.

- **Don't** lose focus or you will become lost in your note-taking.

Copyright © Glencoe/McGraw-Hill, a division of The McGraw-Hill Companies, Inc.

# Using Your Science Notebook

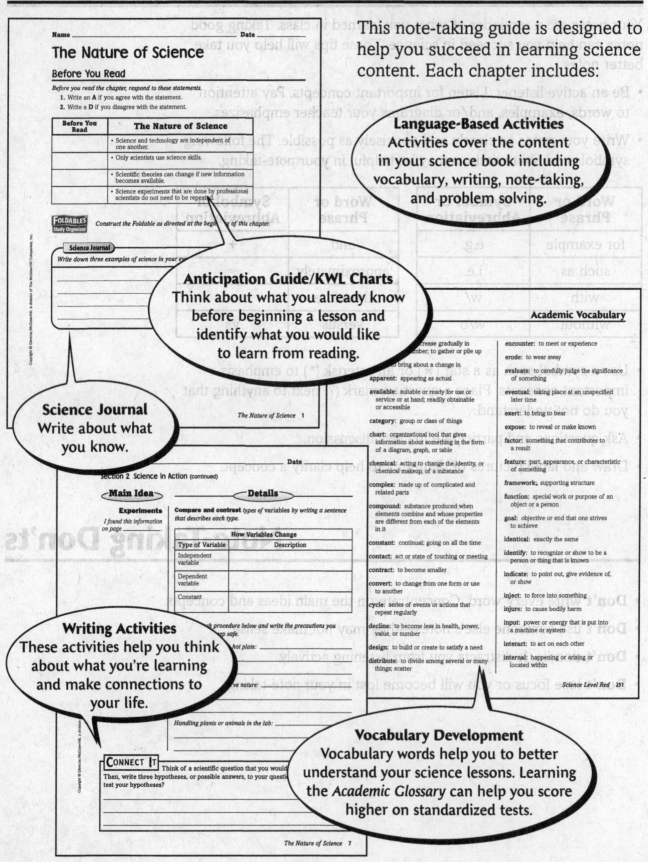

This note-taking guide is designed to help you succeed in learning science content. Each chapter includes:

**Language-Based Activities**
Activities cover the content in your science book including vocabulary, writing, note-taking, and problem solving.

**Anticipation Guide/KWL Charts**
Think about what you already know before beginning a lesson and identify what you would like to learn from reading.

**Science Journal**
Write about what you know.

**Writing Activities**
These activities help you think about what you're learning and make connections to your life.

**Vocabulary Development**
Vocabulary words help you to better understand your science lessons. Learning the *Academic Glossary* can help you score higher on standardized tests.

Copyright © Glencoe/McGraw-Hill, a division of The McGraw-Hill Companies, Inc.

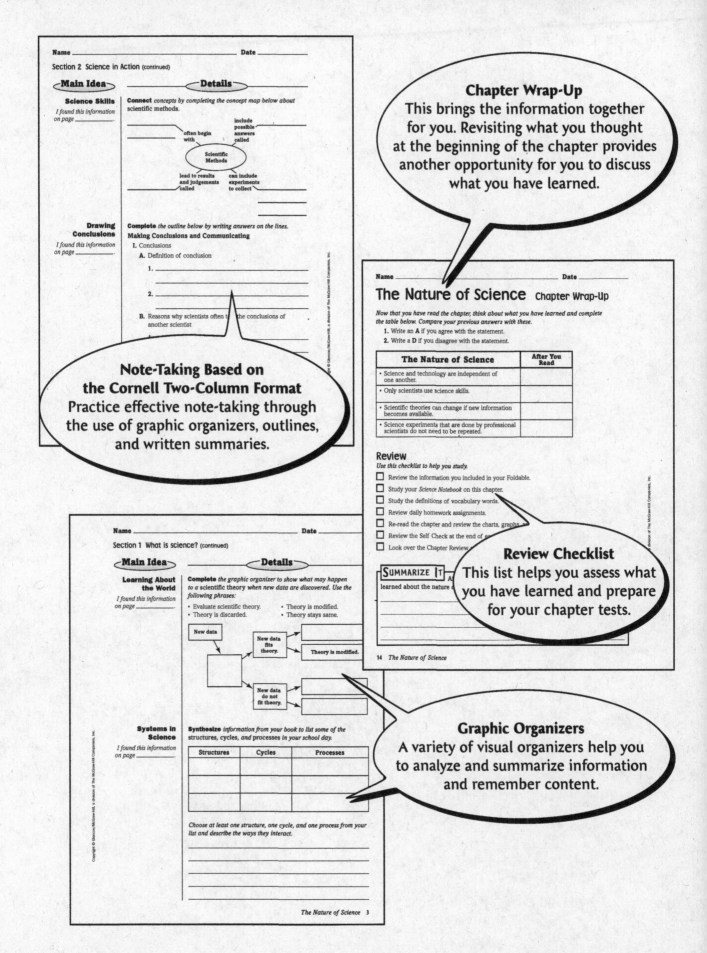

## Chapter Wrap-Up
This brings the information together for you. Revisiting what you thought at the beginning of the chapter provides another opportunity for you to discuss what you have learned.

## Note-Taking Based on the Cornell Two-Column Format
Practice effective note-taking through the use of graphic organizers, outlines, and written summaries.

## Review Checklist
This list helps you assess what you have learned and prepare for your chapter tests.

## Graphic Organizers
A variety of visual organizers help you to analyze and summarize information and remember content.

Copyright © Glencoe/McGraw-Hill, a division of The McGraw-Hill Companies, Inc.

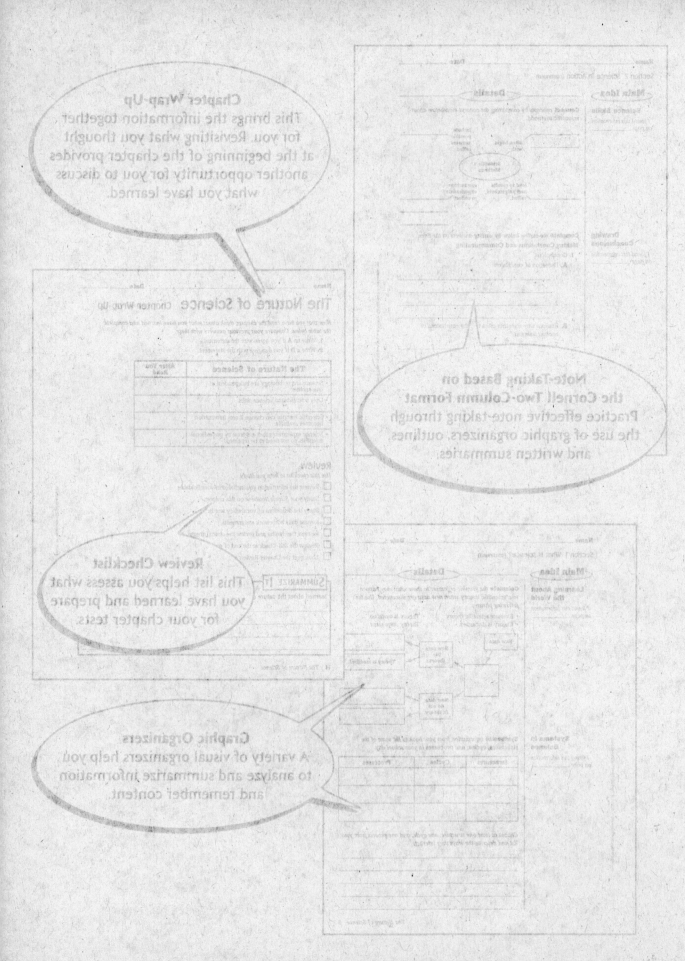

# The Nature of Science

## Before You Read

*Before you read the chapter, respond to these statements.*

1. Write an **A** if you agree with the statement.
2. Write a **D** if you disagree with the statement.

| Before You Read | The Nature of Science |
|---|---|
| | • Science and technology are independent of one another. |
| | • Only scientists use science skills. |
| | • Scientific theories can change if new information becomes available. |
| | • Science experiments that are done by professional scientists do not need to be repeated. |

*Construct the Foldable as directed at the beginning of this chapter.*

**Science Journal**

*Write down three examples of science in your everyday life.*

_____

_____

_____

Copyright © Glencoe/McGraw-Hill, a division of The McGraw-Hill Companies, Inc.

# The Nature of Science
## Section 1 What is science?

**Skim** *through Section 1 of your text. Write three questions that come to mind from reading the headings and looking at the illustrations.*

1. _____

_____

2. _____

_____

3. _____

_____

**Review Vocabulary**

**Define** theory *using your book or a dictionary. Write a sentence about a theory you have heard people talk about in everyday life.*

theory _____

_____

_____

**New Vocabulary**

*Write the correct key term from your text next to each definition.*

_____ an explanation of a pattern observed repeatedly in the natural world

_____ a way of learning more about the natural world

_____ a collection of structures, cycles, and processes that relate to and interact with each other

_____ a rule that describes a pattern in nature

**Academic Vocabulary**

*Use a dictionary to help you write a scientific definition of the word* cycle.

cycle _____

_____

_____

Copyright © Glencoe/McGraw-Hill, a division of The McGraw-Hill Companies, Inc.

## Section 1  What is science? (continued)

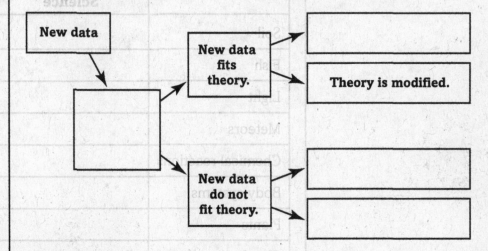

## Main Idea ⟩          ⟨ Details

### Learning About the World

*I found this information on page* _____.

**Complete** *the graphic organizer to show what may happen to a* scientific theory *when new data are discovered. Use the following phrases:*

- Evaluate scientific theory.
- Theory is discarded.
- Theory is modified.
- Theory stays same.

New data

New data fits theory.

Theory is modified.

New data do not fit theory.

### Systems in Science

*I found this information on page* _____.

**Synthesize** *information from your book to list some of the* structures, cycles, *and* processes *in your school day.*

| Structures | Cycles | Processes |
|---|---|---|
|  |  |  |
|  |  |  |
|  |  |  |

*Choose at least one structure, one cycle, and one process from your list and describe the ways they interact.*

_____

_____

_____

_____

Copyright © Glencoe/McGraw-Hill, a division of The McGraw-Hill Companies, Inc.

## Main Idea ─ ─ ─ ─ Details ─

### The Branches of Science

*I found this information on page* _____.

**Classify** *which branch of science—physical science, Earth science, or life science—includes each of the following examples. Then, write one additional example studied by that science.*

| Example | Branch of Science | Additional Example |
|---|---|---|
| Soil | | |
| Fish | | |
| Light | | |
| Meteors | | |
| Chemical reactions | | |
| Body systems | | |
| Plants | | |
| Clouds | | |

### Science and Technology

*I found this information on page* _____.

**Complete** *the following paragraph about the relationship between* science *and* technology.

_____ is a way to learn about the natural world.

To use these answers for helping people, however, they must be

applied in some way. _____ is the practical use of

_____ in our everyday lives.

---

**CONNECT IT** Write about a time that you used science to figure out a problem in your everyday life. Include an additional question about this topic that you might like to investigate.

_____

_____

_____

_____

Copyright © Glencoe/McGraw-Hill, a division of The McGraw-Hill Companies, Inc.

# The Nature of Science

## Section 2  Science in Action

**Skim** *the headings in Section 2. Then make three predictions about what you will learn.*

1. _____

2. _____

3. _____

**Review Vocabulary**

**Define** *observation and give an example of an observation you made today.*

observation

_____

_____

**New Vocabulary**

*Use your book or a dictionary to define the following key terms.*

hypothesis

_____

_____

infer

_____

_____

controlled experiment

_____

_____

variable

_____

_____

constant

_____

_____

**Academic Vocabulary**

*Use a dictionary to define* chart *as it refers to science.*

chart

_____

Copyright © Glencoe/McGraw-Hill, a division of The McGraw-Hill Companies, Inc.

## Section 2 Science in Action (continued)

### Main Idea

#### Science Skills

*I found this information on page _____.*

#### Details

**Connect** *concepts by completing the concept map below about* scientific methods.

_____          include
                         possible
        often begin      answers
        with             called
                 Scientific
                 Methods
        lead to results  can include
        and judgements   experiments
        called           to collect

#### Drawing Conclusions

*I found this information on page _____.*

**Complete** *the outline below by writing answers on the lines.*

**Making Conclusions and Communicating**

**I.** Conclusions

  **A.** Definition of conclusion

   1. _____

      _____

   2. _____

  **B.** Reasons why scientists often test the conclusions of another scientist

   1. _____

      _____

   2. _____

**II.** Communicating—Reasons why it is important for scientists to communicate

  **A.** _____

  **B.** _____

Copyright © Glencoe/McGraw-Hill, a division of The McGraw-Hill Companies, Inc.

## Section 2 Science in Action (continued)

### Main Idea | Details

**Experiments**

*I found this information on page _____.*

**Compare and contrast** *types of* variables *by writing a sentence that describes each type.*

| How Variables Change | |
|---|---|
| Type of Variable | Description |
| Independent variable | |
| Dependent variable | |
| Constant | |

**Laboratory Safety**

*I found this information on page _____.*

**Analyze** *each procedure below and write the precautions you should take to keep safe.*

*Heating a liquid on a hot plate:* _____

_____

_____

*Going outside to observe nature:* _____

_____

_____

*Handling plants or animals in the lab:* _____

_____

**CONNECT IT** Think of a scientific question that you would like to answer. Then, write three hypotheses, or possible answers, to your question. How could you test your hypotheses?

_____

_____

Copyright © Glencoe/McGraw-Hill, a division of The McGraw-Hill Companies, Inc.

# The Nature of Science

## Section 3  Models in Science

**Scan** *Section 3 of your book. Then write three questions that you have about the use of models in science. Try to answer your questions as you read.*

1. _____

2. _____

3. _____

**Review Vocabulary**

**Define** scientific method *using your book or a dictionary. Then give an example of the scientific method in action.*

scientific method _____

_____

**New Vocabulary**

*Use your book or a dictionary to define* model. *Then give some examples of real-life and scientific models.*

model _____

_____

_____

**Academic Vocabulary**

*Use a dictionary to define* encounter. *Then use the term in an original sentence that shows its scientific meaning.*

encounter _____

_____

Copyright © Glencoe/McGraw-Hill, a division of The McGraw-Hill Companies, Inc.

## Section 3  Models in Science (continued)

### Main Idea ─ Details

**Why are models necessary?**

I found this information on page _____.

**Summarize** *how* models *are helpful.*

_____

_____

_____

**Types of Models**

I found this information on page _____.

**Organize** *information in the chart to describe the three types of models and their uses.*

| Models | | |
| --- | --- | --- |
| Type | Description | Use |
| Physical | | |
| | built using computer software | |
| | | help people understand abstract concepts that often are beyond common experience |

**Making Models**

I found this information on page _____.

**Create** *a diagram of the building in which you live. Provide as much detail as possible so that your model will be accurate. Identify uses for this model.*

Copyright © Glencoe/McGraw-Hill, a division of The McGraw-Hill Companies, Inc.

Name _____  Date _____

## Section 3  Models in Science (continued)

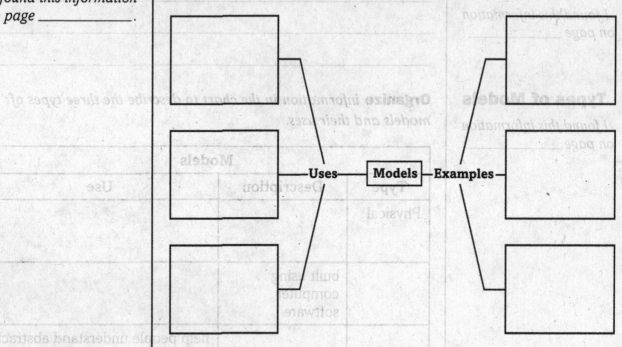

### Main Idea

**Using Models**

I found this information
on page _____.

### Details

**Complete** *the graphic organizer about three ways that models are useful and three examples of scientific models.*

Uses — Models — Examples

**Limitations
of Models**

I found this information
on page _____.

**Identify** *two reasons that models have limitations and list an example of a model for each reason.*

1. _____

_____

_____

2. _____

_____

_____

_____

**CONNECT IT**  As more has been learned about the solar system, the models used to represent it have changed. What are some other models that might have changed over time as new discoveries were made?

_____

_____

Copyright © Glencoe/McGraw-Hill, a division of The McGraw-Hill Companies, Inc.

Name _____ Date _____

# The Nature of Science

## Section 4  Evaluating Scientific Explanation

**Skim** through the section. Read the headings and look at the illustrations. Then write three questions that come to mind. Add to these impressions as you read the section.

1. _____

   _____

2. _____

   _____

3. _____

   _____

**Review Vocabulary**

**Define** prediction *using your book. Write a scientific sentence to give an example of a prediction.*

prediction  _____

   _____

   _____

**New Vocabulary**

*Use your book to define the following terms.*

critical thinking  _____

   _____

data  _____

   _____

**Academic Vocabulary**

*Use* evaluate *in a scientific sentence.*

evaluate  _____

   _____

Copyright © Glencoe/McGraw-Hill, a division of The McGraw-Hill Companies, Inc.

## Section 4 Evaluating Scientific Explanation (continued)

<Main Idea>                    <Details>

**Believe it or not?**

*I found this information on page _____.*

**Complete** *the following sentences using these terms.*

sense      inferences      evaluate      observations

conclusions      accurate      critical

You can _____ an explanation using _____

thinking. First, you should examine the _____ and

decide if you believe they are _____. Then, look at the

_____ or _____ made about the data and

decide if they make _____.

**Evaluating the Data**

*I found this information on page _____.*

**Summarize** *three features of* reliable data.

1. _____

_____

2. _____

_____

3. _____

_____

**Organize** *three characteristics of* good notes.

```
                              _____
                             /
                            /
┌──────────────┐          /
│  Good notes  │— are ───<──────────────────────────
└──────────────┘          \
                           \
                            _____
```

Copyright © Glencoe/McGraw-Hill, a division of The McGraw-Hill Companies, Inc.

## Section 4  Evaluating Scientific Explanation (continued)

| Main Idea | Details |
|---|---|

### Main Idea

**Evaluating the Conclusions**

*I found this information on page* _____ .

### Details

**Complete** *the concept web to show the steps you might use when* **evaluating a** *scientific explanation.* **Use phrases:**

- Are there good notes?
- Could there be another explanation?
- Can the data be repeated?
- Evaluate the conclusion.

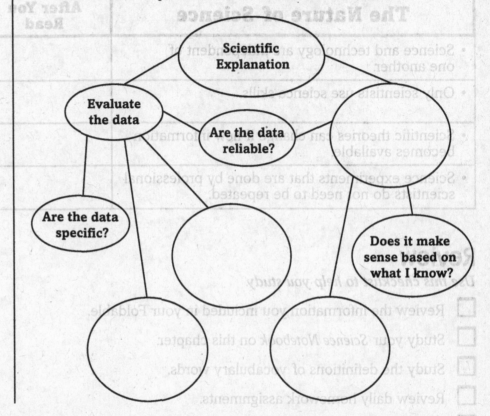

Scientific Explanation

Evaluate the data

Are the data reliable?

Are the data specific?

Does it make sense based on what I know?

---

**CONNECT IT**  Create your own advertisement for a wrinkle cream. Include claims about the product's safety and effectiveness, and use information that might help support those claims. List reasons why another person should or should not believe your ad.

**Advertisement:** _____

_____

_____

**Reasons:** _____

_____

_____

Copyright © Glencoe/McGraw-Hill, a division of The McGraw-Hill Companies, Inc.

# The Nature of Science  Chapter Wrap-Up

*Now that you have read the chapter, think about what you have learned and complete the table below. Compare your previous answers with these.*

1. Write an **A** if you agree with the statement.
2. Write a **D** if you disagree with the statement.

| The Nature of Science | After You Read |
|---|---|
| • Science and technology are independent of one another. | |
| • Only scientists use science skills. | |
| • Scientific theories can change if new information becomes available. | |
| • Science experiments that are done by professional scientists do not need to be repeated. | |

# Review

*Use this checklist to help you study.*

☐ Review the information you included in your Foldable.

☐ Study your *Science Notebook* on this chapter.

☐ Study the definitions of vocabulary words.

☐ Review daily homework assignments.

☐ Re-read the chapter and review the charts, graphs, and illustrations.

☐ Review the Self Check at the end of each section.

☐ Look over the Chapter Review at the end of the chapter.

## SUMMARIZE IT
After reading this chapter, identify three things that you have learned about the nature of scientific investigation.

_____

_____

_____

_____

Copyright © Glencoe/McGraw-Hill, a division of The McGraw-Hill Companies, Inc.

# Measurement

## Before You Read

*Before you read the chapter, respond to these statements.*

1. Write **A** if you agree with the statement.
2. Write **D** if you disagree with the statement.

| Before You Read | Measurement |
|---|---|
|  | • Measurements are recorded by using numbers. |
|  | • Measurements can be precise but not accurate. |
|  | • Most scientists use inches and feet to record length. |
|  | • A bar graph shows parts of a whole. |

*Construct the Foldable as directed at the beginning of the chapter.*

**Science Journal**

*As a member of the pit crew, how can you determine the miles per gallon a car uses? Write in your Science Journal how you would calculate this.*

_____

_____

_____

_____

_____

Copyright © Glencoe/McGraw-Hill, a division of The McGraw-Hill Companies, Inc.

# Measurement

## Section 1  Description and Measurement

**Skim** *Section 1 of your book. Write three questions that come to mind from reading the headings of this section.*

1. _____

2. _____

3. _____

**Review Vocabulary**   **Define** description *to show its scientific meaning.*

description
_____

**New Vocabulary**   *Define each vocabulary term using your book or a dictionary.*

estimation
_____

precision
_____

accuracy
_____

_____

**Academic Vocabulary**   *Use a dictionary to define* significant. *Use* significant *in an original sentence to show its scientific meaning.*

significant
_____

_____

_____

Copyright © Glencoe/McGraw-Hill, a division of The McGraw-Hill Companies, Inc.

Section 1  Description and Measurement (continued)

## Main Idea | Details

**Measurement**

*I found this information on page _____.*

**Define** measurement. *Then give five examples of things that are measured.*

Measurement is _____

Examples:

1. _____
2. _____
3. _____
4. _____
5. _____

**Estimation**

*I found this information on page _____.*

**Distinguish** *two situations in which you might use* estimation.

1. _____
2. _____

**Precision and Accuracy**

*I found this information on page _____.*

**Contrast** precision *and* accuracy. *Define each term. Then complete the Venn diagram with an example of measurements that are* precise, accurate, *and* both precise and accurate.

Precision is _____.

Accuracy is _____.

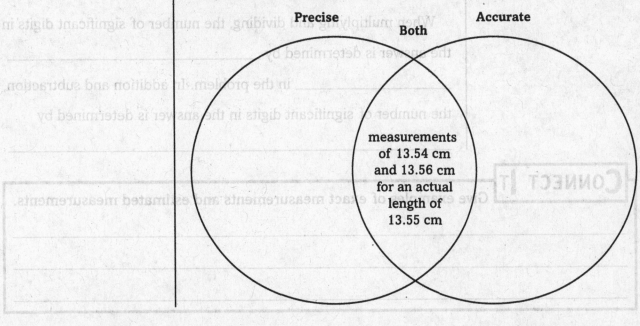

Precise          Accurate

Both

measurements
of 13.54 cm
and 13.56 cm
for an actual
length of
13.55 cm

Copyright © Glencoe/McGraw-Hill, a division of The McGraw-Hill Companies, Inc.

## Section 1 Description and Measurement (continued)

| Main Idea | Details |
|---|---|

### Precision and Accuracy

*I found this information on page _____.*

**Sequence** *the steps to follow when rounding a measurement.*

| Look at the digit _____. |
|---|

| If the digit is less than 5, _____ _____ | If the digit is 5 or greater, _____ _____ |
|---|---|

*I found this information on page _____.*

**Complete** *the chart of rules for using* significant digits. *Identify each category as always, sometimes, or never significant.*

| Type of Digit | Significant? |
|---|---|
| non-zero digits | |
| zeros between other digits | |
| zeros at the beginning of a number | |
| zeros in whole numbers | |

**Summarize** *how to use significant digits in multiplication and division and in addition and subtraction.*

When multiplying and dividing, the number of significant digits in the answer is determined by _____

_____ in the problem. In addition and subtraction, the number of significant digits in the answer is determined by

_____.

### CONNECT IT

**Give examples of exact measurements and estimated measurements.**

_____

_____

_____

Copyright © Glencoe/McGraw-Hill, a division of The McGraw-Hill Companies, Inc.

# Measurement
## Section 2  SI Units

**Predict** *three things you expect to learn in Section 2 after reading its title and headings.*

1. _____

2. _____

3. _____

**Review Vocabulary**   **Define** variable *to show its scientific meaning.*

*variable*   _____

_____

**New Vocabulary**   *Write the correct vocabulary term next to each definition.*

_____   SI unit for mass

_____   amount of change of one measurement in a given amount of time

_____   International System of Units

_____   amount of matter in an object

_____   amount of space an object occupies

_____   SI temperature scale

_____   SI unit for length

_____   measure of the gravitational force on an object

**Academic Vocabulary**   *Use a dictionary to define* summary.

*summary*   _____

_____

Copyright © Glencoe/McGraw-Hill, a division of The McGraw-Hill Companies, Inc.

## Section 2 SI Units (continued)

### Main Idea

### Details

**The International System**

*I found this information on page _____.*

**Sequence** *the prefixes used in the* SI system *from smallest to largest. Write each prefix in the proper place on the diagram.*

centi-　　deka-　　hecto-　　mega-　　milli-

deci-　　giga-　　kilo-　　micro-　　nano-

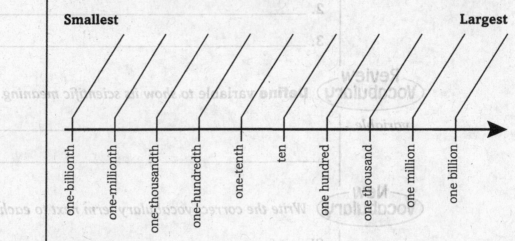

**Smallest**　　　　　　　　　　　　　　　　**Largest**

one-billionth　one-millionth　one-thousandth　one-hundredth　one-tenth　ten　one hundred　one thousand　one million　one billion

**Length**

*I found this information on page _____.*

**Organize** *information about* length *in the graphic organizer.*

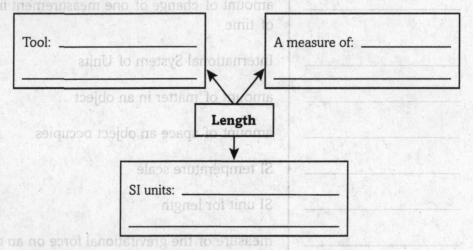

Tool: _____
_____

A measure of: _____
_____

Length

SI units: _____
_____

**Volume**

*I found this information on page _____.*

**Distinguish** *methods of finding* volume.

Regular, square or rectangular objects: _____

_____

Irregular objects: _____

_____

_____

Copyright © Glencoe/McGraw-Hill, a division of The McGraw-Hill Companies, Inc.

## Section 2 SI Units (continued)

### ⟨Main Idea⟩ _____ ⟨Details⟩

**Mass**
*I found this information on page _____.*

**Contrast** mass *and* weight. *Complete the chart.*

|  | Mass | Weight |
|---|---|---|
| What is it a measure of? |  |  |
| What SI units are used to measure it? |  |  |
| Is it the same everywhere? |  |  |

**Temperature**
*I found this information on page _____.*

**Label** *the diagrams to identify important temperatures in the three temperature scales. Circle the scale that is used for SI units.*

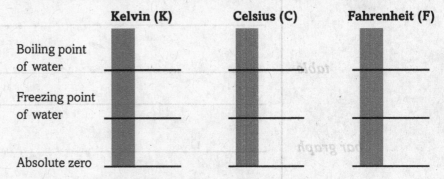

Kelvin (K)    Celsius (C)    Fahrenheit (F)

Boiling point of water

Freezing point of water

Absolute zero

**Time and Rates**
*I found this information on page _____.*

**Analyze** *the relationship between* time *and* rate.

Time is measured in _____ in the SI system. It tells

_____. A rate is _____

_____

_____

### SYNTHESIZE IT

Explain why it is important to have a standard system of units for scientists to use for measuring.

_____

_____

_____

Copyright © Glencoe/McGraw-Hill, a division of The McGraw-Hill Companies, Inc.

# Measurement

## Section 3  Drawings, Tables, and Graphs

**Scan** *Section 3. Write two facts you discovered as you scanned the section.*

1. _____

2. _____

**Review Vocabulary**  *Write an original sentence to show the scientific meaning of model.*

model  _____
_____

**New Vocabulary**  **Define** *each vocabulary term using your book or a dictionary.*

circle graph  _____
_____

table  _____
_____
_____

bar graph  _____
_____

line graph  _____
_____

graph  _____
_____

**Academic Vocabulary**  *Use a dictionary to define category. Use category in an original sentence to show its scientific meaning.*

category  _____
_____
_____

Copyright © Glencoe/McGraw-Hill, a division of The McGraw-Hill Companies, Inc.

## Section 3 Drawings, Tables, and Graphs (continued)

**Main Idea**                                    **Details**

### Scientific Illustrations

*I found this information on page _____.*

**Compare and contrast** drawings, photographs, *and* movies.

| Drawings | Photographs | Movies |
|---|---|---|
|  |  |  |

### Tables and Graphs

*I found this information on page _____.*

**Complete** *the outline to describe* tables *and* graphs.

I. Tables

   **A.** _____

   **B.** _____

II. Graphs

   **A.** _____

   **B.** _____

*I found this information on page _____.*

**Create** *a sample* line graph. *Label the* x-axis *and* y-axis.

**Summarize** *what kind of data can be shown on a line graph.*

_____

_____

Copyright © Glencoe/McGraw-Hill, a division of The McGraw-Hill Companies, Inc.

## Section 3 Drawings, Tables, and Graphs (continued)

### Main Idea

### Details

**Tables and Graphs**

*I found this information on page _____.*

**Model** *a bar graph of your own. Write a caption explaining each part of the graph.*

*I found this information on page _____.*

**Sequence** *the steps to follow to create a* circle graph.

1. _____

2. _____

3. _____

*I found this information on page _____.*

**Evaluate** *why it is important to examine the scale on a graph. Explain why a broken scale is sometimes useful.*

_____

_____

_____

---

**SYNTHESIZE IT**

Compare the two graphs of U.S. endangered species per year in your book. Which do you think is more accurate? Which shows the data most clearly? Why? What other type of graph might you use to show these data?

_____

_____

_____

_____

Copyright © Glencoe/McGraw-Hill, a division of The McGraw-Hill Companies, Inc.

# Tie It Together

*Suppose that you have been asked to design your ideal science classroom. The builder wants to know what measurements will be needed to make your room. Create a plan for your classroom. Include at least one item for which each of the following will need to be measured: length, volume, mass, and temperature. Predict a time measurement for your construction. Suggest ways that each can be measured. Create a scientific illustration showing the design of your room.*

Copyright © Glencoe/McGraw-Hill, a division of The McGraw-Hill Companies, Inc.

# Measurement Chapter Wrap-Up

*Now that you have read the chapter, think about what you have learned and complete the table below. Compare your previous answers with these.*

1. Write an **A** if you agree with the statement.
2. Write **D** if you disagree with the statement.

| Measurement | After You Read |
|---|---|
| • Measurements are recorded by using numbers. | |
| • Measurements can be precise but not accurate. | |
| • Most scientists use inches and feet to measure length. | |
| • A bar graph shows parts of a whole. | |

# Review

*Use this checklist to help you study.*

☐ Review the information you included in your Foldable.

☐ Study your *Science Notebook* on this chapter.

☐ Study the definitions of vocabulary words.

☐ Review daily homework assignments.

☐ Re-read the chapter and review the charts, graphs, and illustrations.

☐ Review the Self Check at the end of each section.

☐ Look over the Chapter Review at the end of the chapter.

## SUMMARIZE IT

After reading this chapter, identify three things that you have learned about measurement.

_____

_____

_____

_____

Copyright © Glencoe/McGraw-Hill, a division of The McGraw-Hill Companies, Inc.

# Matter and Its Changes

## Before You Read

*Preview the chapter title, section titles, and section headings. Complete the chart by listing at least two ideas for each section in each column.*

| K<br>What I know | W<br>What I want to find out |
| --- | --- |
|  |  |
|  |  |
|  |  |
|  |  |

**FOLDABLES**
**Study Organizer**

*Construct the Foldable as directed at the beginning of this chapter.*

**Science Journal**

*Wendy Craig Duncan carried the Olympic flame underwater on the way to the 2000 Summer Olympics in Sydney, Australia. How many different states of matter do you think would be involved in this task? List as many as you can.*

_____

_____

_____

_____

_____

_____

Copyright © Glencoe/McGraw-Hill, a division of The McGraw-Hill Companies, Inc.

# Matter and Its Changes
## Section 1  Physical Properties and Changes

**Scan** *Section 1 of your book. Write a sentence about physical properties of matter.*

| W What I want to find out | K What I know |
|---|---|
| | |

**Review Vocabulary**

**Define** mass *to show its scientific meaning.*

*mass* _____

**New Vocabulary**

*Use your book to write a definition for each word listed below.*

*matter* _____

_____

*physical change* _____

*density* _____

*states of matter* _____

_____

*melting point* _____

_____

*boiling point* _____

_____

**Academic Vocabulary**

*Use your book or a dictionary to define* identify.

*identify* _____

_____

Copyright © Glencoe/McGraw-Hill, a division of The McGraw-Hill Companies, Inc.

Section 1 Physical Properties and Changes (continued)

### Main Idea — Details

**Using Your Senses**

*I found this information on page _____ .*

**Create** *a drawing below to represent the senses you use for making observations. Label each drawing with the sense it represents. Identify those senses that should not be used in the lab.*

**Physical Properties**

*I found this information on page _____ .*

**Complete** *the statement below about physical properties.*

Physical Properties of a material can be _____

_____

Physical Properties you observe include

1. _____     3. _____

2. _____

Physical Properties you can measure include

1. _____     3. _____

2. _____     4. _____

**States of Matter**

*I found this information on page _____ .*

**Sequence** *the four states of matter of any substance according to its temperature by completing the blanks.*

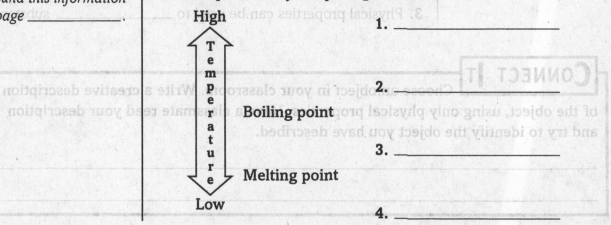

High

Temperature

Boiling point

Melting point

Low

1. _____

2. _____

3. _____

4. _____

Copyright © Glencoe/McGraw-Hill, a division of The McGraw-Hill Companies, Inc.

## Section 1  Physical Properties and Changes (continued)

**Main Idea** ──────── **Details** ────────

### Metallic Properties

*I found this information on page _____.*

**Organize** *the information on metallic properties below. Each circle should include a metallic property and a description of the property. The first one has been done for you.*

**1.** Luster—shine, or how a material reflects light

**2.** _____ can be hammered, pressed, or rolled into sheets

**Metallic Properties**

**3.** Ductility— _____ _____

**4.** _____ _____ _____

### Using Physical Properties

*I found this information on page _____.*

**Summarize** *three ways that you can use the physical properties of substances by completing the blanks in the sentences below.*

**1.** Physical properties can be used to _____ substances.

**2.** Physical properties can be used to _____ substances.

**3.** Physical properties can be used to _____ substances.

**CONNECT IT** Choose an object in your classroom. Write a creative description of the object, using only physical properties. Have a classmate read your description and try to identify the object you have described.

_____

_____

Copyright © Glencoe/McGraw-Hill, a division of The McGraw-Hill Companies, Inc.

# Matter and Its Changes
## Section 2  Chemical Properties and Changes

**Scan** the title and headings in Section 2. Predict three things that might be discussed in this section.

1. _____

2. _____

3. _____

**Review Vocabulary**

**Define** the word heat as it relates to the states of matter. Use your book or a dictionary for help.

heat        _____

**New Vocabulary**

Use each of the words below in an original sentence that reflects the word's scientific meaning.

chemical property  _____

_____

chemical change  _____

_____

law of conservation of mass  _____

_____

**Academic Vocabulary**

Use a dictionary to find the scientific meaning of react.

react        _____

**Ability to Change**

I found this information on page _____ .

**Contrast** physical properties and chemical properties. Write a summary of the differences between these properties.

_____

_____

_____

Copyright © Glencoe/McGraw-Hill, a division of The McGraw-Hill Companies, Inc.

## Section 2  Chemical Properties and Changes (continued)

### Main Idea

**Common Chemical Properties**

*I found this information on page _____.*

### Details

**Complete** *the chart as you read the section. The left column lists common chemical properties. The right column gives an example of that property. The first row of the chart has been done for you.*

| Type of Chemical Property | Example |
|---|---|
| Flammability | Wood will burn. |
| Reacts with oxygen | |
| | Silver can tarnish. |
| | A vitamin can change to another substance. |
| Reacts when heated or cooled | |
| | Water breaks down, or decomposes. |

**Something New**

*I found this information on page _____.*

**Identify** *six signs that a chemical change has occurred.*

1. _____

2. _____

3. _____

4. _____

5. _____

6. _____

Copyright © Glencoe/McGraw-Hill, a division of The McGraw-Hill Companies, Inc.

## Section 2  Chemical Properties and Changes (continued)

**‹Main Idea›**                    **‹Details›**

### Something New

*I found this information on page _____.*

**Compare and contrast** *chemical changes and physical changes by completing the Venn diagram with at least five facts.*

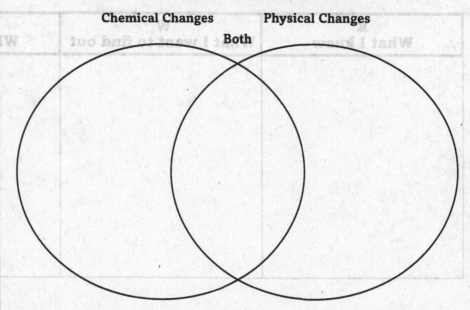

Chemical Changes        Physical Changes
                Both

### The Law of Conservation of Mass

*I found this information on page _____.*

**Create** *a diagram of a campfire below. Label your drawing to show the chemical change that is occurring and how mass is conserved.*

┌─────────────────────────────────────────────────────────────┐

│                                                             │

└─────────────────────────────────────────────────────────────┘

┌──────────────────────────────────────────────────────────────┐
│ **CONNECT IT** │ Give two examples of how understanding chemical properties can be useful in your daily life.
│
│ 1. _____
│    _____
│
│ 2. _____
│    _____
└──────────────────────────────────────────────────────────────┘

Copyright © Glencoe/McGraw-Hill, a division of The McGraw-Hill Companies, Inc.

# Matter and Its Changes Chapter Wrap-Up

*Review the ideas you listed in the chart at the beginning of the chapter. Cross out any incorrect information in the first column. Then complete the chart by filling in the third column.*

| K<br>What I know | W<br>What I want to find out | L<br>What I learned |
| --- | --- | --- |
|  |  |  |

## Review

*Use this checklist to help you study.*

- ☐ Review the information you included in your Foldable.
- ☐ Study your *Science Notebook* on this chapter.
- ☐ Study the definitions of vocabulary words.
- ☐ Review daily homework assignments.
- ☐ Re-read the chapter and review the charts, graphs, and illustrations.
- ☐ Review the Self Check at the end of each section.
- ☐ Look over the Chapter Review at the end of the chapter.

**SUMMARIZE IT** After reading this chapter, identify three things that you have learned about matter and how it changes.

_____

_____

_____

Copyright © Glencoe/McGraw-Hill, a division of The McGraw-Hill Companies, Inc.

# Atoms, Elements, and the Periodic Table

## Before You Read

*Preview the chapter title, section titles, and the section headings. List at least two ideas for each section in each column.*

| K<br>What I know | W<br>What I want to find out |
|---|---|
|  |  |
|  |  |
|  |  |

**FOLDABLES™**
**Study Organizer**

*Construct the Foldable as directed at the beginning of this chapter.*

**Science Journal**

*Make a list of three questions that you think of when you see hot air balloons.*

_____

_____

_____

_____

_____

_____

Copyright © Glencoe/McGraw-Hill, a division of The McGraw-Hill Companies, Inc.

# Atoms, Elements, and the Periodic Table

## Section 1  Structure of Matter

**Read** *the* What You'll Learn *statements for Section 1. Write three questions that come to mind. Look for answers to each question as you read the section.*

1. _____

2. _____

3. _____

**Review Vocabulary**  **Define** density *to show its scientific meaning.*

*density*

_____

**New Vocabulary**  *Write the correct vocabulary word next to each definition.*

_____  small particle that makes up most kinds of matter

_____  uncharged particle in the nucleus of an atom

_____  invisible, negatively charged particle

_____  anything that has mass and takes up space

_____  statement that matter is not created or destroyed, but only changes its form

_____  positively charged central part of an atom

_____  positively charged particle in the nucleus of an atom

**Academic Vocabulary**  *Use a dictionary to define* theory.

*theory*

_____

Copyright © Glencoe/McGraw-Hill, a division of The McGraw-Hill Companies, Inc.

# Atoms, Elements, and the Periodic Table

## Before You Read

*Preview the chapter title, section titles, and the section headings. List at least two ideas for each section in each column.*

| K<br>What I know | W<br>What I want to find out |
| --- | --- |
|  |  |
|  |  |

**Construct the Foldable as directed at the beginning of this chapter.**

**Science Journal**

*Make a list of three questions that you think of when you see hot air balloons.*

_____

_____

_____

_____

_____

_____

Copyright © Glencoe/McGraw-Hill, a division of The McGraw-Hill Companies, Inc.

# Atoms, Elements, and the Periodic Table
## Section 1  Structure of Matter

**Read** *the* What You'll Learn *statements for Section 1. Write three questions that come to mind. Look for answers to each question as you read the section.*

1. _____

2. _____

3. _____

**Review Vocabulary**

**Define** density *to show its scientific meaning.*

*density*

_____

**New Vocabulary**

*Write the correct vocabulary word next to each definition.*

_____  small particle that makes up most kinds of matter

_____  uncharged particle in the nucleus of an atom

_____  invisible, negatively charged particle

_____  anything that has mass and takes up space

_____  statement that matter is not created or destroyed, but only changes its form

_____  positively charged central part of an atom

_____  positively charged particle in the nucleus of an atom

**Academic Vocabulary**

*Use a dictionary to define* theory.

*theory*

_____

Copyright © Glencoe/McGraw-Hill, a division of The McGraw-Hill Companies, Inc.

Section 1  Structure of Matter (continued)

## Main Idea

**Details**

**What is matter? What isn't matter?**

*I found this information on page _____.*

**What makes up matter?**

*I found this information on page _____.*

**State** *the two characteristics common to all* **matter.**

1. _____

2. _____

**Label** *each example as* **matter** *or* **not matter.**

air _____        light _____

heat _____        water _____

**Organize** *Democritus's ideas about* atoms. *Complete the concept map.*

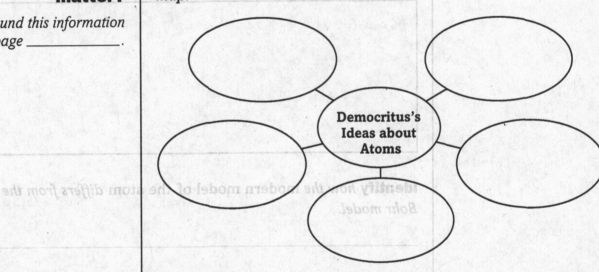

Democritus's Ideas about Atoms

**Identify** *the two main ideas in* Dalton's atomic theory of matter.

1. _____

2. _____

**Summarize** Lavoisier's experiment *and the conclusion he drew from it.*

| Experiment: | Conclusion: |
|---|---|
| | |

Copyright © Glencoe/McGraw-Hill, a division of The McGraw-Hill Companies, Inc.

Name _____     Date _____

⟨ **Main Idea** ⟩                    ⟨ **Details** ⟩

### Models of the Atom

*I found this information on page* _____.

**Compare and contrast** *the* Thomson *and* Rutherford atomic models.

_____

_____

_____

_____

*I found this information on page* _____.

**Create** *a drawing of the* Bohr atom. *Label the* positively charged, negatively charged, *and* neutral parts.

**Identify** *how the* modern model of the atom *differs from the* Bohr model.

_____

_____

**ANALYZE IT**   Make a relative time line of atomic models. List the models from oldest to youngest. State the new discovery that was made with the development of each new model.

Copyright © Glencoe/McGraw-Hill, a division of The McGraw-Hill Companies, Inc.

# Atoms, Elements, and the Periodic Table

## Section 2  The Simplest Matter

**Skim** *the headings and subheadings in Section 2. Write three predictions about what you will learn in this section.*

1. _____

2. _____

3. _____

**Review Vocabulary**

*Write a scientific sentence using the word* **mass.**

*mass*  _____

**New Vocabulary**

*Write the correct vocabulary term next to each definition.*

_____  matter made of only one kind of atom

_____  number of protons in the nucleus of each atom of an element

_____  atom of an element with a different number of neutrons

_____  the number of protons plus the number of neutrons in an atom

_____  weighted average mass of the isotopes of an element

_____  element that generally has a shiny luster and is a good conductor of heat and electricity

_____  element that is usually dull in appearance and is a poor conductor of heat and electricity

_____  element that has characteristics of metals and nonmetals

**Academic Vocabulary**

**Define** unique *using a dictionary.*

*unique*  _____

Copyright © Glencoe/McGraw-Hill, a division of The McGraw-Hill Companies, Inc.

## Section 2 The Simplest Matter (continued)

Main Idea _____ <ellipse>Details</ellipse>

### The Elements

*I found this information on page _____.*

**Summarize** *three key facts about* **elements.**

1. _____

2. _____

3. _____

### The Periodic Table

*I found this information on page _____.*

**Complete** *the graphic organizer to show how the* **periodic table** *is organized.*

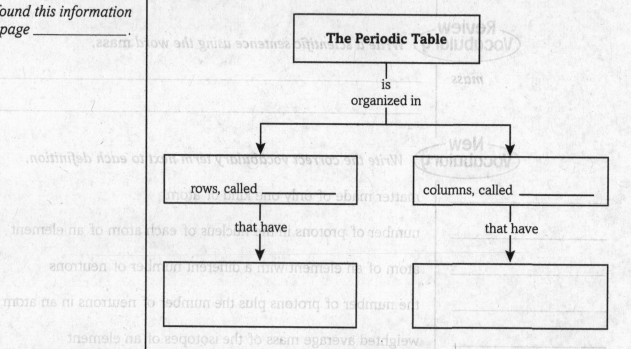

The Periodic Table

is
organized in

rows, called _____

that have

columns, called _____

that have

### Identifying Characteristics

*I found this information on page _____.*

**Label** *the square below with information you would find about* **chlorine** *on the periodic table. Identify each piece of information and explain what you can learn from it.*

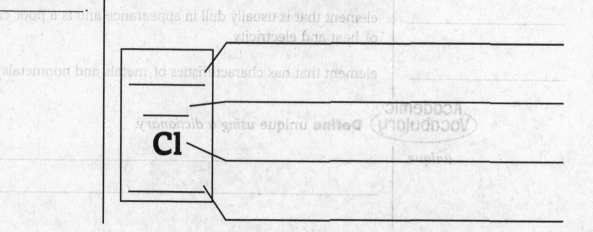

Cl

Copyright © Glencoe/McGraw-Hill, a division of The McGraw-Hill Companies, Inc.

## Section 2  The Simplest Matter (continued)

| Main Idea | Details |
|---|---|

### Identifying Characteristics

*I found this information on page _____ .*

**Contrast** *the three* isotopes *of hydrogen. Complete the chart.*

| Isotope | Protium | Deuterium | Tritium |
|---|---|---|---|
| Number of protons | | | |
| Number of neutrons | | | |
| Mass number | | | |

### Classification of Elements

*I found this information on page _____ .*

**Summarize** *the four characteristics of each type of element in the chart below.*

| | Metals | Nonmetals | Metalloids |
|---|---|---|---|
| 1. | | | |
| 2. | | | |
| 3. | | | |
| 4. | | | |

---

## SYNTHESIZE IT

Metals, nonmetals, and metalloids are located in specific areas of the periodic table. Use what you know about elements and the periodic table to explain why this is.

_____

_____

_____

_____

Copyright © Glencoe/McGraw-Hill, a division of The McGraw-Hill Companies, Inc.

# Atoms, Elements, and the Periodic Table

## Section 3 Compounds and Mixtures

**Scan** *Section 3 using the checklist below.*

☐ Read all section headings.

☐ Read all bold words.

☐ Read all charts and graphs.

☐ Look at the pictures.

☐ Think about what you already know about compounds and mixtures.

*Write two facts you learned about compounds and mixtures as you scanned the section.*

1. _____

_____

2. _____

_____

**Review Vocabulary**

**Define** formula. *Then use the term in an original sentence to show its scientific meaning.*

formula _____

_____

**New Vocabulary**

*Use each vocabulary term in a scientific sentence.*

substance _____

compound _____

mixture _____

**Academic Vocabulary**

*Use a dictionary to define* symbol. *Give an example of a symbol you have used in science.*

symbol _____

_____

Copyright © Glencoe/McGraw-Hill, a division of The McGraw-Hill Companies, Inc.

Section 3 Compounds and Mixtures (continued)

## Main Idea                                    ## Details

**Substances**

*I found this information on page _____.*

**Classify** *the types of substances. Complete the graphic organizer by describing each type and giving two examples.*

```
                    ┌──────────────────┐
                    │    Substances    │
                    └──────────────────┘
                         │       │
              ┌──────────┘       └──────────┐
              ▼                              ▼
```

| Type: _____ | Type: _____ |
| Description: _____ | Description: _____ |
| _____ | _____ |
| _____ | _____ |
| _____ | _____ |
| Examples: _____ | Examples: _____ |
| _____ | _____ |

*I found this information on page _____.*

**Summarize** *what information is contained in the formula of a compound.*

_____

_____

_____

**Analyze** *the formula of each compound. Identify which elements are in each compound and how many atoms of each element make up one unit of the compound.*

|  | **Water** | **Hydrogen peroxide** | **Carbon dioxide** | **Carbon monoxide** |
|---|---|---|---|---|
| Formula | $H_2O$ | $H_2O_2$ | $CO_2$ | $CO$ |
| Atoms and elements |  |  |  |  |

Copyright © Glencoe/McGraw-Hill, a division of The McGraw-Hill Companies, Inc.

Section 3 Compounds and Mixtures (continued)

## Main Idea / Details

**Mixtures**

I found this information on page _____.

**Contrast** compounds *and* mixtures. *Complete the Venn diagram with at least five facts.*

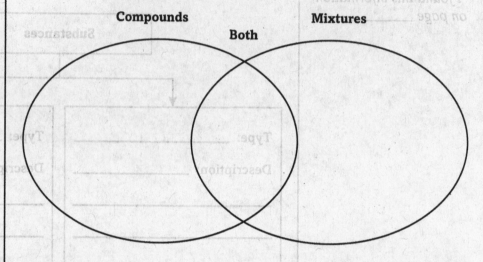

Compounds    Both    Mixtures

I found this information on page _____.

**Summarize** *characteristics of* homogeneous *and* heterogeneous mixtures.

A homogeneous mixture _____.

You _____ see the individual parts. A heterogeneous mixture _____.

You _____ see the individual parts.

Examples of a homogeneous mixture: _____

_____

Examples of a heterogeneous mixture: _____

_____

**CONNECT IT** Give example of two mixtures and two compounds that are important to your everyday life.

_____

_____

Copyright © Glencoe/McGraw-Hill, a division of The McGraw-Hill Companies, Inc.

# Tie It Together

*The formulas for three substances are listed below.*

- *Describe the properties of each substance as thoroughly as you can.*
- *Identify each as an element or a compound.*
- *Write the number of protons in the nuclei of the element or elements in each substance.*
- *State whether those elements are metals, nonmetals, or metalloids, and any properties you can infer for those elements.*
- *Use a periodic table.*

1. Water ($H_2O$): _____

_____

_____

_____

_____

2. Table salt (NaCl): _____

_____

_____

_____

_____

3. Gold (Au): _____

_____

_____

_____

_____

Copyright © Glencoe/McGraw-Hill, a division of The McGraw-Hill Companies, Inc.

# Atoms, Elements, and the Periodic Table  Chapter Wrap-Up

*Review the ideas you listed in the chart at the beginning of the chapter. Cross out any incorrect information in the first column. Then complete the chart by filling in the third column. How do your ideas now compare with those you provided at the beginning of the chapter?*

| **K**<br>**What I know** | **W**<br>**What I want to find out** | **L**<br>**What I learned** |
| --- | --- | --- |
|  |  |  |
|  |  |  |
|  |  |  |

# Review

*Use this checklist to help you study.*

☐ Review the information you included in your Foldable.

☐ Study your *Science Notebook* on this chapter.

☐ Study the definitions of vocabulary words.

☐ Review daily homework assignments.

☐ Re-read the chapter and review the charts, graphs, and illustrations.

☐ Review the Self Check at the end of each section.

☐ Look over the Chapter Review at the end of the chapter.

## SUMMARIZE IT
After reading this chapter, identify three things that you have learned about atoms and elements.

_____

_____

_____

_____

Copyright © Glencoe/McGraw-Hill, a division of The McGraw-Hill Companies, Inc.

# Motion, Forces, and Simple Machines

## Before You Read

*Before you read the chapter, think about what you know about these topics. List three things that you already know about motion, forces, and simple machines in the first column. Then list three things that you would like to learn about these topics in the second column.*

| K<br>What I know | W<br>What I want to find out |
|---|---|
|  |  |
|  |  |
|  |  |

Construct the Foldable as directed at the beginning of this chapter.

**Science Journal**

*Write a paragraph comparing the motion of a ball and a paper airplane being thrown high in the air and returning to the ground.*

_____

_____

_____

_____

_____

_____

Copyright © Glencoe/McGraw-Hill, a division of The McGraw-Hill Companies, Inc.

# Motion, Forces, and Simple Machines

## Section 1  Motion

**Scan** *the headings in Section 1 of your book. Identify three topics that will be discussed.*

1. _____

2. _____

3. _____

**Review Vocabulary**

**Define** meter *using your book or a dictionary.*

meter

_____

_____

**New Vocabulary**

*Use your book to define the following terms.*

average speed

_____

_____

instantaneous speed

_____

_____

velocity

_____

_____

_____

acceleration

_____

_____

_____

_____

**Academic Vocabulary**

*Use a dictionary to define* exert *to show its scientific meaning.*

exert

_____

Copyright © Glencoe/McGraw-Hill, a division of The McGraw-Hill Companies, Inc.

## Section 1  Motion (continued)

**Main Idea**                    **Details**

**Speed**

I found this information
on page _____.

**Skim** *the section, and create a graphic organizer that identifies three different ways speed can be described.*

I found this information
on page _____.

**Complete** *the equations to show how to calculate average speed and distance.*

### Calculating Average Speed

$$\text{speed (in m/s)} = \frac{\boxed{\phantom{xxx}} \text{(in m)}}{\boxed{\phantom{xxx}} \text{(in s)}} \qquad s = \frac{\boxed{\phantom{x}}}{\boxed{\phantom{x}}}$$

### Calculating Distance Traveled

$$\text{distance traveled (in m)} = \boxed{\phantom{xxx}} \text{(in m/s)} \times \boxed{\phantom{xxx}} \text{(in s)}$$

$$d = \boxed{\phantom{xxx}}$$

**Velocity**

I found this information
on page _____.

**Identify** *the factors that affect velocity.*

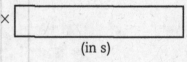

Velocity changes when

Copyright © Glencoe/McGraw-Hill, a division of The McGraw-Hill Companies, Inc.

**Section 1  Motion** (continued)

## ⬭Main Idea⬭          ⬭Details⬭

### Acceleration

*I found this information on page* _____.

**Complete** *the equations to show how to calculate the* acceleration *of an object that changes speed but not direction.*

**Calculating Acceleration**

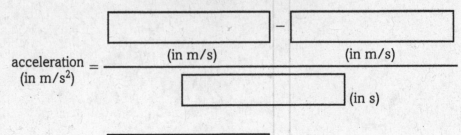

$$\text{acceleration} \atop \text{(in m/s}^2) = \frac{\boxed{\phantom{xxxxxxxx}} \ \text{(in m/s)} \ - \ \boxed{\phantom{xxxxxxxx}} \ \text{(in m/s)}}{\boxed{\phantom{xxxxxxxx}} \ \text{(in s)}}$$

$$a = \frac{\boxed{\phantom{xxxx}} - \boxed{\phantom{xxxx}}}{\boxed{\phantom{xxxxxxxx}}}$$

*I found this information on page* _____.

**Compare** *changes in the speed of an object by identifying what is happening to the speed during each segment of the graph.*

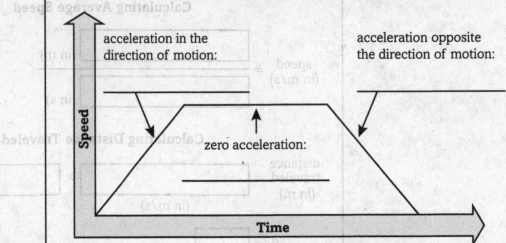

acceleration in the direction of motion: _____

zero acceleration: _____

acceleration opposite the direction of motion: _____

**C**ONNECT **I**T   Explain how you can use your watch and a car's odometer to determine the average speed of a trip by car.

_____

_____

_____

_____

Copyright © Glencoe/McGraw-Hill, a division of The McGraw-Hill Companies, Inc.

# Motion, Forces, and Simple Machines

## Section 2  Newton's Laws of Motion

**Skim** *Section 2 of your book. Write three questions that come to mind. Look for answers to your questions as you read the section.*

1. _____

2. _____

3. _____

**Review Vocabulary**

**Define** gravity *using your book or a dictionary.*

gravity

_____

_____

**New Vocabulary**

*Use your book to define the following terms. Then use each term in a sentence to show its scientific meaning.*

force

_____

_____

_____

Newton's laws of motion

_____

_____

friction

_____

_____

inertia

_____

_____

**Academic Vocabulary**

*Use a dictionary to define* constant.

constant

_____

Copyright © Glencoe/McGraw-Hill, a division of The McGraw-Hill Companies, Inc.

## Section 2 Newton's Laws of Motion (continued)

### ⟨Main Idea⟩

### ⟨Details⟩

**Force**

*I found this information on page _____.*

**Model** *two ways unbalanced forces can be combined. Use arrows labeled as* force 1, force 2, *and* net force *to indicate sizes and directions of force. Then draw a model to represent balanced forces on an object.*

| Unbalanced Forces | Balanced Forces |
|---|---|
|  |  |

**Newton's Laws of Motion** and **Newton's First Law**

*I found this information on page _____.*

**Organize** *information about* Newton's first law *by completing the chart below.*

| Newton's First Law | | |
|---|---|---|
| **If an object is:** | **Then:** | **Unless:** |
| at rest |  | a net force is applied to it |
| in motion |  |  |

**Newton's Second Law**

*I found this information on page _____.*

**Summarize** Newton's second law *in your own words. Then complete the equation used to calculate acceleration.*

_____

_____

**Newton's Second Law**

$$\text{acceleration} \atop \text{(in meters/second}^2) = \frac{\boxed{\phantom{xxxxxx}} \text{(in newtons)}}{\boxed{\phantom{xxxxxx}} \text{(in kilograms)}}$$

Copyright © Glencoe/McGraw-Hill, a division of The McGraw-Hill Companies, Inc.

Section 2  Newton's Laws of Motion (continued)

## Main Idea

*I found this information on page _____.*

**Summarize** *the relationship between mass, inertia, and acceleration by completing the blanks.*

As an object's mass increases, its inertia _____, and

acceleration requires _____ force.

As an object's mass decreases, its inertia _____, and

acceleration requires _____ force.

### Newton's Third Law

*I found this information on page _____.*

**Model** *how* action and reaction forces *act in pairs. Draw a situation in which a force pair acts on different objects. Use arrows to label the* action *and* reaction *forces. Below your drawing, explain how the forces act and how the motions of the objects change.*

_____

_____

_____

_____

**CONNECT IT**  Describe why the equal forces involved in Newton's third law of motion are not considered balanced forces.

_____

_____

_____

Copyright © Glencoe/McGraw-Hill, a division of The McGraw-Hill Companies, Inc.

# Motion, Forces, and Simple Machines

Section 3  Work and Simple Machines

**Scan** *Section 3 of your book. Write three things that you want to learn about work and simple machines.*

1. _____

2. _____

3. _____

### Review Vocabulary

**Define** radius *using your book or a dictionary.*

radius _____

### New Vocabulary

*Use your book to define the following terms. Then write a paragraph using the terms.*

work _____

_____

simple machine _____

_____

compound machine _____

_____

mechanical advantage _____

_____

_____

_____

### Academic Vocabulary

*Use a dictionary to define* input *to show its scientific meaning.*

input _____

Copyright © Glencoe/McGraw-Hill, a division of The McGraw-Hill Companies, Inc.

## Section 3  Work and Simple Machines (continued)

| Main Idea | Details |
|---|---|

### Work

*I found this information on page _____.*

**Define** *the two things that must happen for work to be done by completing the graphic organizer.*

For work to be done

### Calculating Work

*I found this information on page _____.*

**Complete** *the equations for calculating work.*

**Calculating Work**

$$\text{work (in J)} = \boxed{\phantom{xxxxxxxx}} \times \boxed{\phantom{xxxxxxxx}}$$
$$\text{(in N)} \qquad \text{(in m)}$$

$$\boxed{\phantom{xx}} = \boxed{\phantom{xx}}$$

### What is a machine?

*I found this information on page _____.*

**Summarize** *two ways a machine can make work easier.*

Machines make work easier by changing:

1. _____

_____

2. _____

_____

### The Pulley

*I found this information on page _____.*

**Compare** *pulleys by completing the chart.*

| Pulleys | | |
|---|---|---|
| **Type of Pulley System** | **Effect on Force** | **Mechanical Advantage** |
| Single pulley | | |
| Double-pulley system | | |

Copyright © Glencoe/McGraw-Hill, a division of The McGraw-Hill Companies, Inc.

## Section 3  Work and Simple Machines (continued)

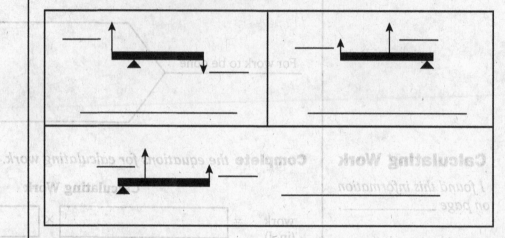

### Main Idea

### Details

#### The Lever

*I found this information on page _____.*

**Label** *the arrows on each of the diagrams below as either input force* ($F_i$) *or output force* ($F_o$). *Then identify the class of lever that each diagram represents.*

#### The Inclined Plane

*I found this information on page _____.*

**Compare** *how the amount of force needed to move an object changes with the length of the inclined plane. Complete the blanks below with* less, more, *or* the most.

**Longer Inclined Plane:**

_____ force is needed to move an object

**Shorter Inclined Plane:**

_____ force is needed to move an object

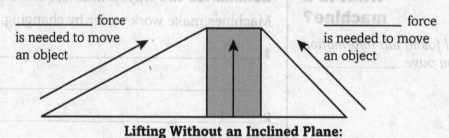

**Lifting Without an Inclined Plane:**

_____ force is needed to move the object

*I found this information on page _____.*

**Complete** *the graphic organizer to identify three examples of inclined planes.*

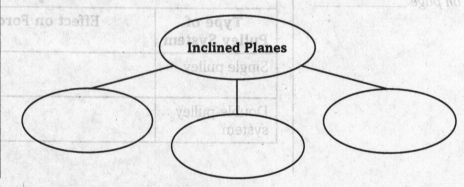

Inclined Planes

Copyright © Glencoe/McGraw-Hill, a division of The McGraw-Hill Companies, Inc.

Name _____  Date _____

# Tie It Together

## Synthesize

*You will be loading heavy crates into a truck. The crates are too heavy to lift up to the bed of the truck. What simple machines could you use to help you, and how would you use them? Make use of as many as you can. Write about them and draw diagrams to show how they will help you.*

Copyright © Glencoe/McGraw-Hill, a division of The McGraw-Hill Companies, Inc.

# Motion, Forces, and Simple Machines
## Chapter Wrap-Up

*Review the ideas you listed in the chart at the beginning of the chapter. Cross out any incorrect information in the first column. Then complete the chart by filling in the third column.*

| K<br>What I know | W<br>What I want to find out | L<br>What I learned |
|---|---|---|
|  |  |  |

## Review

*Use this checklist to help you study.*

- ☐ Review the information you included in your Foldable.
- ☐ Study your *Science Notebook* on this chapter.
- ☐ Study the definitions of vocabulary words.
- ☐ Review daily homework assignments.
- ☐ Re-read the chapter and review the charts, graphs, and illustrations.
- ☐ Review the Self Check at the end of each section.
- ☐ Look over the Chapter Review at the end of the chapter.

**SUMMARIZE IT** After reading this chapter, identify three main ideas you learned that you did not know before.

_____

_____

_____

_____

Copyright © Glencoe/McGraw-Hill, a division of The McGraw-Hill Companies, Inc.

# Energy

## Before You Read

*Before you read the chapter, respond to these statements.*

1. Write an **A** if you agree with the statement.
2. Write a **D** if you disagree with the statement.

| Before You Read | Energy |
|---|---|
| | • A moving object has energy. |
| | • A generator creates new energy. |
| | • Temperature is a form of energy. |
| | • Chemical reactions can give off energy. |

**FOLDABLES** **Study Organizer**

*Construct the Foldable as directed at the beginning of this chapter.*

**Science Journal**

*List three changes that you have seen occur today, and describe what changed.*

_____

_____

_____

_____

_____

_____

_____

Copyright © Glencoe/McGraw-Hill, a division of The McGraw-Hill Companies, Inc.

# Energy

## Section 1  Energy Changes

**Scan** the headings in Section 1 of your book. Then, write four questions about energy. Try to answer your questions as you read.

1. _____

2. _____

3. _____

4. _____

**Review Vocabulary**

Use the term speed *in a sentence that shows its scientific meaning.*

*speed* _____

_____

**New Vocabulary**

**Define** *each vocabulary term using your book or a dictionary.*

*energy* _____

_____

*kinetic energy* _____

_____

*potential energy* _____

_____

*law of conservation of energy* _____

_____

**Academic Vocabulary**

*Use a dictionary to define* transform.

*transform* _____

_____

Copyright © Glencoe/McGraw-Hill, a division of The McGraw-Hill Companies, Inc.

## Section 1  Energy Changes (continued)

### Main Idea       Details

**Energy**

*I found this information on page _____.*

**Identify** *four changes caused by energy. Use your book to help you.*

1. _____
   _____

2. _____
   _____

3. _____
   _____

4. _____
   _____

**Forms of Energy**

*I found this information on page _____.*

**Organize** *some familiar energy transformations by completing the chart below.*

| Energy Transformations | | |
| --- | --- | --- |
| Energy Begins as | Where the Change Takes Place | Energy Becomes |
|  | muscles in your body |  |
|  | hot sand at the beach |  |
|  | hands rubbing together |  |
|  | lightbulb |  |

**Kinetic Energy**

*I found this information on page _____.*

**Compare** *the effects of mass and speed on kinetic energy by filling in the blanks below with the terms* **more** *or* **less.**

A moving object with *more* mass has _____ kinetic energy.

A moving object with *less* mass has _____ kinetic energy.

A moving object moving with _____ speed has *more* kinetic energy.

A moving object moving with _____ speed has *less* kinetic energy.

Copyright © Glencoe/McGraw-Hill, a division of The McGraw-Hill Companies, Inc.

## Section 1 Energy Changes (continued)

**Main Idea**                **Details**

### Potential Energy

*I found this information on page _____.*

**Create** *a diagram in the space below that shows the effect of position and gravity on potential and kinetic energy. If you need help, refer to the picture of a ski slope in your book. Be sure to show the following points in your diagram:*

- where kinetic energy is greatest and least
- where potential energy is greatest and least
- where potential energy is increasing
- where kinetic energy is increasing

### Conservation of Energy

*I found this information on page _____.*

**Summarize** *the principles of the law of conservation of energy by completing the following paragraph.*

The total amount of energy in the universe never _____.

This means that energy cannot be _____ or _____.

Energy can, however, change from one _____ to another.

One example of energy changing _____ is when the

_____ energy of water behind a dam is converted into the

_____ energy that spins a generator. The generator converts

this energy into _____ energy and heat. During this process,

the total amount of energy does not _____.

Copyright © Glencoe/McGraw-Hill, a division of The McGraw-Hill Companies, Inc.

# Energy
## Section 2  Temperature

**Scan** *Section 2 of your book using the checklist below.*

☐ Read all section titles.

☐ Read all boldface words.

☐ Look at all of the pictures.

☐ Think about what you already know about temperature.

*Write three facts that you discovered about temperature and heat as you scanned the section.*

1. _____

2. _____

3. _____

**Define** *the following terms by writing the term next to its definition.*

_____  particle formed when two or more atoms bond together

_____

_____  measure of the average kinetic energy of the particles in an object

_____  transfer of energy from one object to another as a result of a difference in temperature

_____  transfer of energy by collisions between atoms in a material

_____  transfer of heat that occurs when particles move between objects or areas that differ in temperature

_____  the transfer of energy by waves

*Use a dictionary to write the scientific definition for* transfer.

*transfer* _____

_____

Copyright © Glencoe/McGraw-Hill, a division of The McGraw-Hill Companies, Inc.

Section 2  Temperature (continued)

## Main Idea

## Details

### Temperature

*I found this information
on page _____ .*

**Analyze** *the effect that temperature has on the speed of motion
and kinetic energy of the molecules of a gas by completing the
chart below.*

| Molecules in a Gas | | |
|---|---|---|
| Temperature | Speed of Motion | Kinetic Energy |
| Low | | |
| High | | |

### Measuring Temperature

*I found this information
on page _____ .*

**Compare** *the Fahrenheit and Celsius temperature scales by
drawing a thermometer below and indicating water's boiling point
and freezing point on each scale.*

### Heat and Temperature

*I found this information
on page _____ .*

**Complete** *the sentences below about temperature increase.*

During summer, the water in a lake generally is _____

_____ . During winter, lake

water generally is _____ .

This temperature difference occurs because _____

_____ or

_____ . Water

absorbs a large amount of heat for each degree of temperature

_____ . Once the lake is warm, it must lose a

_____ .

Copyright © Glencoe/McGraw-Hill, a division of The McGraw-Hill Companies, Inc.

## Section 2 Temperature (continued)

### ⟨Main Idea⟩

**Heat on the Move**

*I found this information on page* _____ .

### ⟨Details⟩

**Complete** *the concept map about the three methods of heat transfer.*

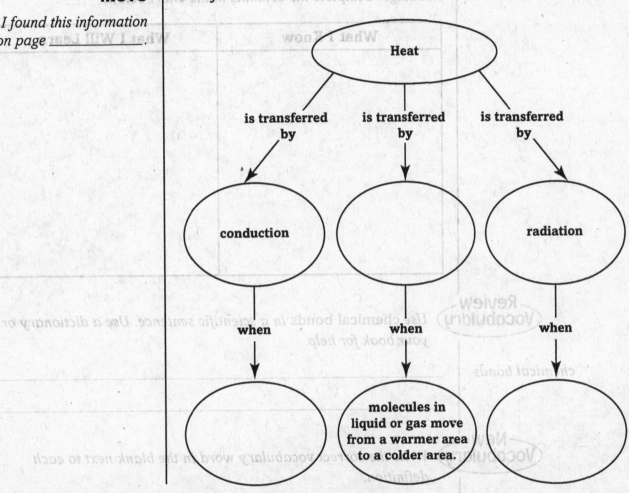

Heat

is transferred by | is transferred by | is transferred by

conduction | | radiation

when | when | when

| | molecules in liquid or gas move from a warmer area to a colder area. |

---

**CONNECT IT**  Imagine yourself stirring a hot cup of hot chocolate with a metal spoon. As you stir, you observe that the spoon becomes hot. Use what you've learned about heat to explain why this happens. In your explanation, describe the method or methods of heat transfer involved.

_____

_____

_____

_____

Copyright © Glencoe/McGraw-Hill, a division of The McGraw-Hill Companies, Inc.

# Energy
## Section 3  Chemical Energy

**Predict** *what you will learn in this section by scanning the headings. Complete the columns in the chart below.*

| What I Know | What I Will Learn |
|---|---|
|  |  |
|  |  |
|  |  |
|  |  |

**Review Vocabulary**  *Use* chemical bonds *in a scientific sentence. Use a dictionary or your book for help.*

chemical bonds   _____

_____

**New Vocabulary**  *Write the correct vocabulary word in the blank next to each definition.*

_____  something that changes the rate of a chemical reaction without being changed itself

_____  a chemical reaction that gives off heat energy

_____  a chemical reaction that absorbs heat energy

**Academic Vocabulary**  *Use the word* compound *in a scientific sentence.*

compound   _____

_____

Copyright © Glencoe/McGraw-Hill, a division of The McGraw-Hill Companies, Inc.

## Section 3 Chemical Energy (continued)

<big>**Main Idea**</big>     <big>**Details**</big>

| | |
|---|---|
| **Chemical Reactions and Energy**<br><br>*I found this information on page _____.* | **List** *three examples of chemical reactions listed in your book.*<br><br>1. _____<br><br>2. _____<br><br>3. _____ |
| **Energy in Reactions**<br><br>*I found this information on page _____.* | **Draw** *a model of the chemical reactions that take place during photosynthesis. Refer to the figure in your book if you need help. Make sure to include these terms:* **carbon dioxide, oxygen, chlorophyll, sunlight, sugar, water.** |

<br><br><br><br><br><br><br><br>

*I found this information on page _____.*

**Compare** *endothermic and exothermic reactions in the Venn diagram below. Use the following terms:*

- require energy
- combustion
- release energy
- photosynthesis
- chemical reaction
- energy is transferred

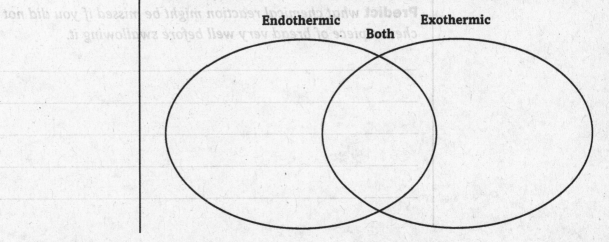

Endothermic     **Both**     Exothermic

Copyright © Glencoe/McGraw-Hill, a division of The McGraw-Hill Companies, Inc.

## Section 3  Chemical Energy (continued)

### Main Idea

### Details

**Energy in Reactions**

I found this information on page _____ .

**Classify** *the following reactions as* exothermic *or* endothermic. *Explain your reasoning.*

candle burning _____

_____

bread baking _____

_____

photosynthesis _____

_____

fireworks exploding _____

_____

I found this information on page _____ .

**Identify** *two ways to increase the rate of a chemical reaction.*

1. _____

2. _____

**Summarize** *the characteristics of a catalyst by completing the following sentence.*

A catalyst _____

_____

**Predict** *what chemical reaction might be missed if you did not chew a piece of bread very well before swallowing it.*

_____

_____

_____

_____

_____

Copyright © Glencoe/McGraw-Hill, a division of The McGraw-Hill Companies, Inc.

# Tie It Together

## Synthesize

*Use your knowledge of energy and your imagination to describe energy transformations in three different situations. Use the topics suggested below, or choose your own.*

### Kinetic and Potential Energy Transformations

• Going on a roller coaster ride

*I've just ridden the roller coaster at the amusement park. As the coaster moved up the*

*first hill, its potential energy increased, reaching its highest* _____

_____

_____

_____

### Heat Energy

• Turning up the thermostat on a winter day

*When I got home from school, the house felt cold, so I turned up the thermostat. After a*

*few minutes, I felt warmer, because* _____

_____

_____

_____

### Chemical Energy

• Roasting marshmallows over a campfire

_____

_____

_____

_____

Copyright © Glencoe/McGraw-Hill, a division of The McGraw-Hill Companies, Inc.

# Energy  Chapter Wrap-Up

*Now that you have read the chapter, think about what you have learned and complete the table below. Compare your previous answers with these.*

  1. Write an **A** if you agree with the statement.
  2. Write a **D** if you disagree with the statement.

| Energy | After You Read |
|---|---|
| • A moving object has energy. | |
| • A generator creates new energy. | |
| • Temperature is a form of energy. | |
| • Chemical reactions can give off energy. | |

# Review

*Use this checklist to help you study.*

- ☐ Review the information you included in your Foldable.
- ☐ Study your *Science Notebook* on this chapter.
- ☐ Study the definitions of vocabulary words.
- ☐ Review daily homework assignments.
- ☐ Re-read the chapter and review the charts, graphs, and illustrations.
- ☐ Review the Self Check at the end of each section.
- ☐ Look over the Chapter Review at the end of the chapter.

**SUMMARIZE IT** After reading this chapter, identify three main ideas that you have learned about energy.

_____

_____

_____

_____

_____

Copyright © Glencoe/McGraw-Hill, a division of The McGraw-Hill Companies, Inc.

# Electricity and Magnetism

## Before You Read

*Preview the chapter title, the section titles, and the section headings. List at least one thing you know and one thing you want to find out for each section of the chapter.*

| K<br>What I know | W<br>What I want to find out |
|---|---|
|  |  |
|  |  |
|  |  |
|  |  |

**FOLDABLES™**
**Study Organizer**

*Construct the Foldable as directed at the beginning of this chapter.*

**Science Journal**

*List five electrical devices you used today and describe what each device did.*

_____

_____

_____

_____

_____

_____

Copyright © Glencoe/McGraw-Hill, a division of The McGraw-Hill Companies, Inc.

# Electricity and Magnetism

## Section 1 Electric Charge and Forces

**Objectives** *Review the section objectives. Write three questions that these statements bring to mind.*

1. _____

2. _____

3. _____

**Review Vocabulary** **Define** atom *to show its scientific meaning.*

atom

_____

_____

**New Vocabulary** *Use your book or a dictionary to define the key terms.*

charging by contact _____

charging by induction _____

static charge _____

_____

electric discharge _____

_____

**Academic Vocabulary** *Use a dictionary to define* contact.

contact _____

_____

Copyright © Glencoe/McGraw-Hill, a division of The McGraw-Hill Companies, Inc.

## Section 1 Electric Charge and Forces (continued)

| ⟨**Main Idea**⟩ | ⟨**Details**⟩ |
| --- | --- |

**Electric Charges**

*I found this information on page _____.*

**Identify** *the parts of the atom in the chart below.*

| Particles That Make Up Atoms | | |
| --- | --- | --- |
| **Particle** | **Charge of Particle** | **Particle Location** |
| Proton | | |
| | | nucleus |
| | negative | |

*I found this information on page _____.*

**Complete** *the statements to determine when atoms have electric charge.*

Atoms have positive charge ⟶ when _____

_____ ⟶ when there are equal numbers of electrons and protons.

Atoms have negative charge ⟶ when _____

**The Forces Between Charges**

*I found this information on page _____.*

**Model** *the forces between like and unlike charges. Draw pictures to show the forces for each situation.*

| Positive Particle/ Negative Particle | Positive Particle/ Positive Particle | Negative Particle/ Negative Particle |
| --- | --- | --- |
| | | |

**Compose** *a sentence describing how electric force depends on distance and on charge.*

Sentence: _____

_____

Copyright © Glencoe/McGraw-Hill, a division of The McGraw-Hill Companies, Inc.

Section 1 Electric Charge and Forces (continued)

**Main Idea**                    **Details**

**Making Objects Electrically Charged**

*I found this information on page _____.*

**Identify and define** *the two ways objects become electrically charged by completing the graphic organizer.*

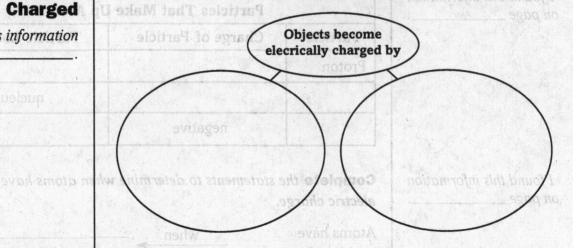

**Conductors and Insulators**

*I found this information on page _____.*

**Organize** *information about insulators and conductors in the chart below.*

| Insulator | Conductor |
| --- | --- |
| What is it? | What is it? |
| Examples: | Examples: |

**CONNECT IT** Make a diagram to show at least four people in a lightning storm. Show some of them acting safely, and some acting unsafely. Use information from the section to explain why each behavior is safe or unsafe.

Copyright © Glencoe/McGraw-Hill, a division of The McGraw-Hill Companies, Inc.

# Electricity and Magnetism
## Section 2 Electric Current

**Scan** *Use the checklist below to preview Section 2 of your book.*

☐ Read all section titles.

☐ Read all bold words.

☐ Look at all of the pictures, charts, and graphs.

☐ Think about what you already know about electric current.

*Write three facts that you discovered about electric current as you scanned the section.*

1. _____

2. _____

3. _____

**Review Vocabulary** *Use* kinetic energy *in a scientific sentence.*

kinetic energy   _____

_____

**New Vocabulary** *Read the definitions below. Write the key term on the blank in the left column.*

_____   measure of how difficult it is for electrons to flow in an object

_____   setup of devices that allows current to follow one closed path

_____   the flow of electric charges

_____   a measure of the amount of electrical energy transferred by an electric charge as it moves from one point to another in a circuit

_____   a setup of devices that allows current to follow more than one closed path

_____   a closed path in which electric charges can flow

**Academic Vocabulary** **Define** *the word* neutral *using a dictionary.*

neutral   _____

Copyright © Glencoe/McGraw-Hill, a division of The McGraw-Hill Companies, Inc.

## Section 2 Electric Current (continued)

<Main Idea>  ———————  <Details>

**Electric Current**

*I found this information on page _____.*

**Complete** *the sentences about electric current.*

Electric current is _____.

Electric current is measured using an SI unit called _____.

**A Simple Electric Circuit**

*I found this information on page _____.*

**Create** *a drawing of a circuit that performs a useful function.*

[ drawing box ]

*List two important facts about how a circuit works.*

_____

_____

_____

_____

**Making Electric Charges Flow**

*I found this information on page _____.*

**Connect** *how each factor affects electric charges in a circuit.*

| Term | How It Affects Electric Charges |
|---|---|
| Electric field | |
| Electric resistance | |
| Battery | |

Copyright © Glencoe/McGraw-Hill, a division of The McGraw-Hill Companies, Inc.

Section 2  Electric Current (continued)

## Main Idea

### Transferring Electrical Energy and Voltage

*I found this information on page _____.*

### Series and Parallel Circuits

*I found this information on page _____.*

## Details

**Define** *Ohm's law by explaining the meaning of each letter in the equation:* V = IR.

V _____ = I _____ x R _____

**Design** *a parallel circuit that has three paths, a battery, and three lightbulbs. Use your book to help you.*

• Label each device.

• Use arrows to show the direction in which electricity flows in each path.

**CONNECT IT**  One bulb in a strand of decorative lights burns out and the rest of the strand stops working. Identify the type of circuit that was used to connect the lights.

_____

_____

_____

Copyright © Glencoe/McGraw-Hill, a division of The McGraw-Hill Companies, Inc.

# Electricity and Magnetism
## Section 3  Magnetism

**Predict**  *Read the title of Section 3. Predict three concepts that might be discussed in this section.*

1. _____

2. _____

3. _____

### Review Vocabulary

*Use* mechanical energy *in a sentence that shows its meaning.*

mechanical energy

_____

_____

_____

### New Vocabulary

*Use the following key terms in original sentences that show their meaning.*

magnetic domain

_____

_____

electromagnet

_____

_____

electromagnetic induction

_____

_____

### Academic Vocabulary

**Define** temporary *using a dictionary. Then use it in a sentence that reflects its scientific meaning.*

temporary

_____

_____

Copyright © Glencoe/McGraw-Hill, a division of The McGraw-Hill Companies, Inc.

## Section 3  Magnetism (continued)

<Main Idea>                    <Details>

**Magnets**

*I found this information on page _____.*

**Model** *how magnets exert forces on each other in the boxes below. Use the figure in your book to help you.*

• Label the poles of the magnets.

• Use arrows to show how the magnets exert forces on each other.

| Two South Poles | North Pole and South Pole | Two North Poles |
|---|---|---|
| | | |
| | | |
| | | |

*Write a general statement about attraction and repulsion of magnets.*

_____

_____

**Magnetic Materials**

*I found this information on page _____.*

**Compare and contrast** *the way that paper clips interact with a magnet and the way paper clips interact with one another by filling in the blanks below.*

_____ of a paper clip do

not normally all point in the same direction. Therefore, paper clips

_____ to one another. The _____

of a magnet mostly point in the _____ direction.

When a magnet is brought near a paper clip, the magnetic domains

of the paper clip _____ so that _____

_____. This causes the paper

clip to be attracted to the magnet.

Copyright © Glencoe/McGraw-Hill, a division of The McGraw-Hill Companies, Inc.

Section 3 Magnetism (continued)

## Main Idea — Details

**Electro-magnetism**

*I found this information on page _____*

**Analyze** *the way electromagnets work by completing the chart. Use your book to help you.*

| Cause | Effect |
|---|---|
| Increasing the current of an electromagnet | |
| | The north and south poles of the magnet will change positions. |

**Generating Electric Current**

*I found this information on page _____.*

**Sequence** *steps to generate electricity by electromagnetic induction.*

| Electricity is generated using the following process: | |
|---|---|
| 1. | |
| 2. | |
| 3. | |

## SYNTHESIZE IT

Suppose that you are given two iron nails, wire, and two batteries of your choice. Draw and label designs for two electromagnets of different strengths made of these materials.

Copyright © Glencoe/McGraw-Hill, a division of The McGraw-Hill Companies, Inc.

# Tie It Together

## Synthesize It

*Identify five everyday devices that work by using electricity. Describe the energy transformations that take place within each device.*

Device _____

_____

_____

Device _____

_____

_____

Device _____

_____

_____

_____

Device _____

_____

_____

Device _____

_____

_____

_____

Copyright © Glencoe/McGraw-Hill, a division of The McGraw-Hill Companies, Inc.

# Electricity and Magnetism
## Chapter Wrap-Up

*Review the ideas you listed in the chart at the beginning of the chapter. Cross out
any incorrect information in the first column. Then complete the chart by filling in
the third column.*

| K<br>What I know | W<br>What I want to find out | L<br>What I learned |
|---|---|---|
| | | |
| | | |
| | | |

# Review
*Use this checklist to help you study.*

- ☐ Review the information you included in your Foldable.
- ☐ Study your *Science Notebook* on this chapter.
- ☐ Study the definitions of vocabulary words.
- ☐ Review daily homework assignments.
- ☐ Re-read the chapter and review the charts, graphs, and illustrations.
- ☐ Review the Self Check at the end of each section.
- ☐ Look over the Chapter Review at the end of the chapter.

> ## SUMMARIZE IT
> After reading this chapter, identify three things that you have
> learned about electricity and magnetism.
> _____
> _____
> _____
> _____

Copyright © Glencoe/McGraw-Hill, a division of The McGraw-Hill Companies, Inc.

# Waves

## Before You Read

*Before you read the chapter, read each statement below.*

**1.** Write an **A** if you agree with the statement.

**2.** Write a **D** if you disagree with the statement.

| Before You Read | Waves |
|---|---|
| | • Waves carry matter and energy. |
| | • There is more than one kind of wave. |
| | • Waves carry different amounts of energy. |
| | • All waves travel at the same speed. |

**FOLDABLES™**
**Study Organizer**

*Construct the Foldable as directed at the beginning of this chapter.*

**Science Journal**

*Write a paragraph about some places where you have seen water waves.*

_____

_____

_____

_____

_____

Copyright © Glencoe/McGraw-Hill, a division of The McGraw-Hill Companies, Inc.

# Waves

## Section 1  What are waves?

Copyright © Glencoe/McGraw-Hill, a division of The McGraw-Hill Companies, Inc.

**Skim** *the title and headings of Section 1. List two things that might be discussed in this section.*

1. _____

2. _____

**Review Vocabulary**  **Define** energy *in your own words.*

energy  _____

_____

_____

**New Vocabulary**  *Define each vocabulary term using your book or a dictionary.*

wave  _____

_____

mechanical wave  _____

_____

transverse wave  _____

_____

_____

compressional wave  _____

_____

electromagnetic wave  _____

_____

_____

**Academic Vocabulary**  *Use a dictionary to define* medium *in its scientific sense.*

medium  _____

_____

_____

**Section 1  What are waves?** (continued)

**Main Idea**  **Details**

### What is a wave?

*I found this information on page _____.*

**Identify** *two types of* waves *that carry energy.*

1. _____

2. _____

*I found this information on page _____.*

**Contrast** *the energy carried in a sound wave and the energy in a moving ball.*

_____

_____

_____

### A Model for Waves

*I found this information on page _____.*

**Create** *your own model for a wave. Use information from your book to make a drawing that models how a wave can move energy without moving matter.*

- Label the parts of your drawing that represent matter and energy.
- Write a caption to explain your drawing.

| My Model for Waves |
|---|
|  |
|  |
| _____ |
| _____ |
| _____ |

Copyright © Glencoe/McGraw-Hill, a division of The McGraw-Hill Companies, Inc.

Section 1  What are waves? (continued)

⟨ **Main Idea** ⟩          ⟨ **Details** ⟩

### Mechanical Waves

*I found this information on page* _____.

**Organize** *information from the section in the outline below.*

Mechanical waves—Travel through a _____.

   **A.** Types of wave mediums

      **1.** _____

      **2.** _____

      **3.** _____

   **B.** Types of Mechanical Waves

      **1.** _____

      **2.** _____

### Making Sound Waves and Electromagnetic Waves

*I found this information on page* _____.

**Compare and contrast** *the characteristics of* sound waves *and* electromagnetic waves *by completing the Venn diagram below.*

- carry energy
- carry radiant energy
- do not need a medium
- mechanical waves
- move through a medium

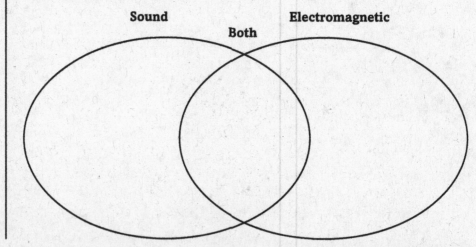

Sound      Both      Electromagnetic

**CONNECT IT**  Evaluate how electromagnetic and mechanical waves are useful in your daily life.

_____

_____

_____

Copyright © Glencoe/McGraw-Hill, a division of The McGraw-Hill Companies, Inc.

# Waves

## Section 2  Wave Properties

**Scan** *Section 2 of your book. Write three facts you discovered about wave properties as you scanned the section.*

1. _____

   _____

2. _____

   _____

3. _____

   _____

**Review Vocabulary**  **Define** *each key term using your book or a dictionary.*

speed _____

**New Vocabulary**

amplitude _____

wavelength _____

frequency _____

**Academic Vocabulary**  *Use the word* **parallel** *in a scientific sentence.*

parallel _____

_____

Copyright © Glencoe/McGraw-Hill, a division of The McGraw-Hill Companies, Inc.

## Section 2 Wave Properties (continued)

### ◁ Main Idea ▷ _____ ◁ Details ▷

**Amplitude**

I found this information on page _____.

**Create** a transverse wave in the space below. Label the crest, trough, and amplitude of the wave on your drawing.

**Wavelength**

I found this information on page _____.

**Complete** the descriptions for determining wavelength of two types of waves in the chart below.

| Wavelength is the distance: | Type of Wave | |
|---|---|---|
| | Transverse | Compressional |
| from one | | |
| to the next | | |
| or from one | | |
| to the next | | |

**Frequency**

I found this information on page _____.

**Model** the relationship between frequency and wavelength when wave speed is the same. In the top box, draw a wave with a frequency of one wavelength per second. In the bottom box, draw a wave with a frequency of two wavelengths per second.

Copyright © Glencoe/McGraw-Hill, a division of The McGraw-Hill Companies, Inc.

## Section 2 Wave Properties (continued)

| Main Idea | Details |
|---|---|

### Wave Speed

*I found this information on page _____.*

**Summarize** *how to use the wave speed equation to calculate wave speed by completing the steps below.*

1. The wave speed equation is

   **wave speed in m/s =**

   _____ × _____

2. To calculate the speed of a wave that has a frequency of 550 Hz and a wavelength of 0.8 m, insert the values into the wave speed equation.

   **wave speed = _____ × _____**

3. Multiply to find the answer.

   Answer: _____

**Compare** *the speeds of different types of waves in different mediums by completing the chart below with the words* gases, liquids, *or* solids.

| How mediums affect wave speed | | |
|---|---|---|
| Wave type | move fastest through | move slowest through |
| mechanical waves | | |
| electromagnetic waves | | |

**CONNECT IT** Individual members of a choir sing at different pitches. Analyze the wavelengths of the sound waves produced by soprano, alto, and baritone singers.

_____

_____

_____

_____

Copyright © Glencoe/McGraw-Hill, a division of The McGraw-Hill Companies, Inc.

# Waves
## Section 3  Wave Behavior

**Predict** *by reading the title and subheadings three things that might be discussed in this section.*

1. _____

2. _____

3. _____

**Review Vocabulary**  *Write a sentence using the word* echo *to reflect its scientific use.*

echo

_____

_____

**New Vocabulary**  *Use the new vocabulary terms to write your own original scientific sentences.*

reflection

_____

refraction

_____

diffraction

_____

_____

interference

_____

_____

**Academic Vocabulary**  **Define** overlap *using a dictionary.*

overlap

_____

_____

_____

Copyright © Glencoe/McGraw-Hill, a division of The McGraw-Hill Companies, Inc.

Section 3 Wave Behavior (continued)

## Main Idea

## Details

### Reflection

I found this information on page _____.

**Skim** the section about reflection. *In the Question spaces, write two questions you have about reflection. As you read the section, write answers to your questions.*

Question: _____

Answer: _____

_____

Question: _____

Answer: _____

_____

### Refraction

I found this information on page _____.

**Create** a diagram below showing what happens to a light wave *as it passes from water to air. Draw a second picture showing what happens as light passes from air to water. Label the* normal *and the* light ray's direction of travel *in each drawing.*

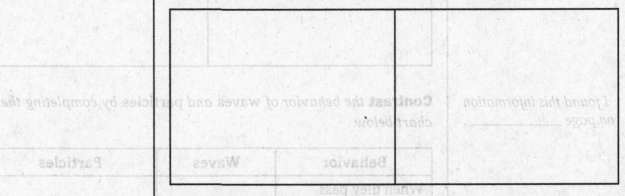

**Summarize** why light refracts *when it passes from one material to another.*

_____

**Sequence** the seven colors into which sunlight separates when it *passes through a prism.*

Longest                    Wavelength                    Shortest

←———————————————————————————————————————————→

| | | | | | | |
|---|---|---|---|---|---|---|

Copyright © Glencoe/McGraw-Hill, a division of The McGraw-Hill Companies, Inc.

## Section 3 Wave Behavior (continued)

**Main Idea**          **Details**

### Diffraction

*I found this information
on page _____.*

**Summarize** *two factors that affect how much a wave can
be diffracted.*

1. _____

2. _____

### What happens when waves meet?

*I found this information
on page _____.*

**Model** constructive *and* destructive interference *in the two boxes
below. Label the crests and troughs of the waves in your model.*

| Interference | |
|---|---|
| **Constructive** | **Destructive** |
| | |

*I found this information
on page _____.*

**Contrast** *the behavior of* waves *and* particles *by completing the
chart below.*

| Behavior | Waves | Particles |
|---|---|---|
| When they pass through an opening | | |
| When they meet | | |

**CONNECT IT**   Use what you have learned about the behavior of waves to
evaluate two ways to protect your ears from damage due to loud noises.

_____

_____

_____

Copyright © Glencoe/McGraw-Hill, a division of The McGraw-Hill Companies, Inc.

# Tie It Together

## Model Wave Motion

*Design a model you could use to study the behavior and properties of waves. Draw your model below.*

**Answer each question about your model.**

**1.** What medium does your model use?

_____

_____

**2.** How could you measure the wavelength of the waves in your model?

_____

_____

**3.** How could you use your model to demonstrate reflection, refraction, and diffraction of waves?

_____

_____

_____

_____

Copyright © Glencoe/McGraw-Hill, a division of The McGraw-Hill Companies, Inc.

# Waves Chapter Wrap-Up

*Now that you have read the chapter, think about what you have learned and complete
the table below. Compare your previous answers with these.*

1. Write an **A** if you agree with the statement.
2. Write a **D** if you disagree with the statement.

| Waves | After You Read |
|---|---|
| • Waves carry matter and energy. | |
| • There is more than one kind of wave. | |
| • Waves carry different amounts of energy | |
| • All waves travel at the same speed. | |

# Review

*Use this checklist to help you study.*

☐ Review the information you included in your Foldable.

☐ Study your *Science Notebook* on this chapter.

☐ Study the definitions of vocabulary words.

☐ Review daily homework assignments.

☐ Re-read the chapter and review the charts, graphs, and illustrations.

☐ Review the Self Check at the end of each section.

☐ Look over the Chapter Review at the end of the chapter.

## SUMMARIZE IT

After reading this chapter, identify three things that you have
learned about waves.

_____

_____

_____

_____

Copyright © Glencoe/McGraw-Hill, a division of The McGraw-Hill Companies, Inc.

# Rocks and Minerals

## Before You Read

*Before you read the chapter, respond to these statements.*

1. Write an **A** if you agree with the statement.
2. Write a **D** if you disagree with the statement.

| Before You Read | Rocks and Minerals |
|---|---|
| | • Minerals are made by people. |
| | • Most rocks consist of one or more minerals. |
| | • Rocks are classified in three major groups. |
| | • Rocks have stopped forming on Earth. |
| | • Rocks and minerals have many uses in society. |

**FOLDABLES™**
**Study Organizer**

*Construct the Foldable as directed at the beginning of this chapter.*

**Science Journal**

*Observe a rock or mineral sample. Write three characteristics about it.*

_____

_____

_____

_____

_____

Copyright © Glencoe/McGraw-Hill, a division of The McGraw-Hill Companies, Inc.

# Rocks and Minerals

## Section 1  Minerals—Earth's Jewels

**Scan** *Section 1 of your book. Then, write three questions that you have about minerals. Try to answer your questions as you read.*

1. _____

2. _____

3. _____

**Review Vocabulary**

**Define** physical property *with the help of your book or a dictionary.*

*physical property*

_____

_____

_____

**New Vocabulary**

*Write the correct vocabulary word from your book next to each definition.*

_____ a solid material that has an orderly, repeating pattern of atoms

_____ a mineral that contains enough of a useful substance that it can be mined at a profit

_____ a rare, valuable mineral that can be cut and polished to give it a beautiful appearance

_____ a solid that is usually made up of two or more minerals

**Academic Vocabulary**

*Use a dictionary to find the definition of* refine *as it applies to metals. Write the definition below in your own words.*

*refine*

_____

_____

_____

Copyright © Glencoe/McGraw-Hill, a division of The McGraw-Hill Companies, Inc.

## Section 1  Minerals—Earth's Jewels (continued)

| Main Idea | Details | Details |

### What is a mineral?

*I found this information on page* _____.

**Complete** *the chart below about minerals.*

| Minerals |
|---|
| Definition: |
| Examples: |
| Ways minerals form: 1. |
| 2. |
| 3. |

### Properties of Minerals

*I found this information on page* _____.

**Contrast** *cleavage and fracture by writing three different characterisitcs of each in the following chart.*

| Cleavage | Fracture |
|---|---|
|  |  |
|  |  |
|  |  |

*I found this information on page* _____.

**Contrast** *the qualities of mineral color and luster.*

Color _____

_____

_____

Luster _____

_____

_____

Copyright © Glencoe/McGraw-Hill, a division of The McGraw-Hill Companies, Inc.

Section 1 Minerals—Earth's Jewels (continued)

**Main Idea**      **Details**

| Common Minerals | Sequence *four steps that describe how copper ore is turned into useful products. The first step has been completed for you.* |
|---|---|

*I found this information on page _____ .*

1. Copper ore is mined and crushed.

2. _____

3. _____

4. _____

*I found this information on page _____ .*

**List** *characteristics of a gem and an ore in the chart below.*

| Gem | Ore |
|---|---|
| | |
| | |
| | |
| | |

**CONNECT IT** Write the names of six objects in your classroom that are made using minerals. Then explain how minerals are important in your everyday life.

1. _____  2. _____  3. _____

4. _____  5. _____  6. _____

_____

_____

_____

_____

Copyright © Glencoe/McGraw-Hill, a division of The McGraw-Hill Companies, Inc.

# Rocks and Minerals
## Section 2  Igneous and Sedimentary Rocks

**Skim** *the headings in Section 2. Then make three predictions about what you will learn.*

1. _____

2. _____

3. _____

**Review Vocabulary**   **Define** *the following terms using your book or a dictionary.*

lava _____

_____

**New Vocabulary**

igneous rock _____

_____

extrusive _____

_____

intrusive _____

_____

sedimentary rock _____

_____

**Academic Vocabulary**

process _____

_____

Copyright © Glencoe/McGraw-Hill, a division of The McGraw-Hill Companies, Inc.

## Section 2  Igneous and Sedimentary Rocks (continued)

### Main Idea | Details

**Igneous Rocks**

*I found this information on page _____.*

**Contrast** *extrusive and intrusive igneous rocks in the chart.*

| Igneous Rocks | | | |
|---|---|---|---|
| Type | Form from molten rock called | Have cooling rate that is | Have crystal size that is |
| Extrusive | | | |
| Intrusive | | | |

*I found this information on page _____.*

**Organize** *a concept map about igneous rocks using these words and phrases.*

- high silica content
- granitic

- low silica content
- dark colored

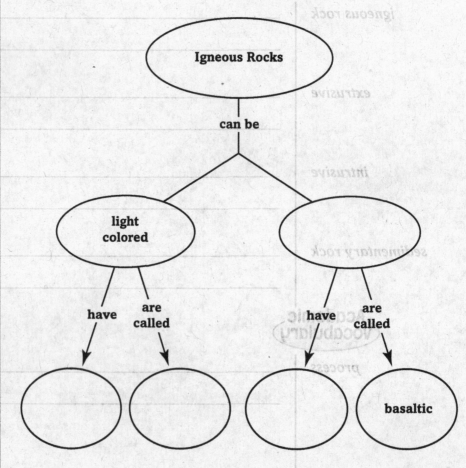

Copyright © Glencoe/McGraw-Hill, a division of The McGraw-Hill Companies, Inc.

## Section 2  Igneous and Sedimentary Rocks (continued)

### Main Idea

**Sedimentary Rocks**

*I found this information on page _____ .*

### Details

**Classify** *sedimentary rocks by some of their characteristics.*

|  | Detrital | Chemical | Organic |
|---|---|---|---|
| Form from |  |  |  |
| How form |  |  |  |
| Where form |  |  |  |
| Examples |  |  |  |

**CONNECT IT** Choose a sedimentary or igneous rock. You might pick basalt, granite, shale, or sandstone. Write a story from the rock's perspective about how the rock formed. When writing your story, you should pretend that you are the rock.

_____

_____

_____

_____

_____

_____

Copyright © Glencoe/McGraw-Hill, a division of The McGraw-Hill Companies, Inc.

# Rocks and Minerals

## Section 3  Metamorphic Rocks and the Rock Cycle

**Scan** the headings in Section 3. Write three predictions about what you will learn in this section.

1. _____

2. _____

3. _____

### Review Vocabulary

**Define** each vocabulary word. Then, write a sentence reflecting the scientific meaning of each of the words.

*pressure* _____

_____

### New Vocabulary

*metamorphic rock* _____

_____

*foliated* _____

_____

*nonfoliated* _____

_____

*rock cycle* _____

_____

### Academic Vocabulary

*layer* _____

_____

Copyright © Glencoe/McGraw-Hill, a division of The McGraw-Hill Companies, Inc.

## Section 3 Metamorphic Rocks and the Rock Cycle (continued)

**Main Idea** | **Details**

### New Rock from Old

*I found this information on page _____.*

**Summarize** *the conditions under which rocks experience metamorphism as you complete the chart below.*

| Conditions of Metamorphic Rock Formation | |
|---|---|
| temperature | |
| pressure | |
| time | |

*I found this information on page _____.*

**Draw** *a metamorphic rock with a foliated texture and a metamorphic rock with a nonfoliated texture below. Show and label two characteristics of each type of rock in the top boxes, and list an example of each type in the bottom boxes.*

| Foliated texture | Nonfoliated texture |
|---|---|
| | |
| Examples: | Examples: |

CONNECT IT

Copyright © Glencoe/McGraw-Hill, a division of The McGraw-Hill Companies, Inc.

## Section 3  Metamorphic Rocks and the Rock Cycle (continued)

| Main Idea | Details | Details |
|---|---|---|

**Rock Cycle**

*I found this information on page* _____.

**Create** *a diagram of the rock cycle below.*
- Label each type of rock in your diagram.
- Label the processes in your diagram. Use the words *melting, cooling, weathering and erosion, compaction and cementation,* and *heat and pressure.*

**Identify** *two other cycles that occur in nature.*

1. _____

2. _____

**CONNECT IT**  While on a leisurely hike, a geologist from the nearby university noticed that the gravel in a sedimentary rock consists of pieces of both igneous and metamorphic rock. As the geologist, write a brief report explaining how this is possible.

_____

_____

_____

Copyright © Glencoe/McGraw-Hill, a division of The McGraw-Hill Companies, Inc.

# Tie It Together

## Design

*Some artists specialize in making art from rock and mineral pieces. The different colors, textures, and other properties of the rocks and minerals can produce spectacular displays. In the space below, design your own rock and mineral art. It might be mounted on a wall, make up the courtyard of a building, or be a large monument. You may use any rock or mineral shown in your book in your art.*

Copyright © Glencoe/McGraw-Hill, a division of The McGraw-Hill Companies, Inc.

# Rocks and Minerals Chapter Wrap-Up

*Now that you have read the chapter, think about what you have learned and complete the table below. Compare your previous answers with these.*

1. Write an **A** if you agree with the statement.
2. Write a **D** if you disagree with the statement.

| Rocks and Minerals | After You Read |
|---|---|
| • Minerals are made by people. | |
| • Most rocks consist of one or more minerals. | |
| • Rocks are classified in three major groups. | |
| • Rocks have stopped forming on Earth. | |
| • Rocks and minerals have many uses in society. | |

# Review

*Use this checklist to help you study.*

☐ Review the information you included in your Foldable.

☐ Study your *Science Notebook* on this chapter.

☐ Study the definitions of vocabulary words.

☐ Review daily homework assignments.

☐ Re-read the chapter and review the charts, graphs, and illustrations.

☐ Review the Self Check at the end of each section.

☐ Look over the Chapter Review at the end of the chapter.

## SUMMARIZE IT
After reading this chapter, identify three things that you have learned about rocks and minerals.

_____

_____

_____

_____

Copyright © Glencoe/McGraw-Hill, a division of The McGraw-Hill Companies, Inc.

# Forces Shaping Earth

## Before You Read

*Preview the chapter title, the section titles, and the section headings. List at least two ideas for each section in each column.*

| K<br>What I know | W<br>What I want to find out |
|---|---|
|  |  |
|  |  |
|  |  |
|  |  |

**FOLDABLES**
**Study Organizer**

**Construct the Foldable as directed at the beginning of this chapter.**

**Science Journal**

*Use descriptive adjectives to describe mountains in a short paragraph.*

_____

_____

_____

_____

_____

_____

_____

Copyright © Glencoe/McGraw-Hill, a division of The McGraw-Hill Companies, Inc.

# Forces Shaping Earth
## Section 1 Earth's Moving Plates

**Scan** *the section before you begin to read.*

☐ Read all section headings.

☐ Read all bold words, highlighted in yellow.

☐ Look at all of the pictures.

*Write three facts that you discovered about Earth's moving plates.*

1. _____

2. _____

3. _____

**Review Vocabulary**   **Define** magma *to show its scientific meaning.*

magma   _____

_____

**New Vocabulary**   *Write the vocabulary term that matches each definition.*

_____   solid, innermost layer of Earth's interior

_____   layer of Earth that lies above the inner core and is thought to be made up mostly of molten metal

_____   largest layer of Earth's interior

_____   Earth's outermost layer

_____   rigid layer of Earth made of the crust and a part of the upper mantle

_____   section of Earth's crust and rigid upper mantle

_____   large fracture in rock along which movement occurs

_____   type of plate movement that occurs when one plate sinks beneath another plate

**Academic Vocabulary**   *Use* contract *in a sentence to reflect its scientific meaning.*

contract   _____

_____

Copyright © Glencoe/McGraw-Hill, a division of The McGraw-Hill Companies, Inc.

## Section 1 Earth's Moving Plates (continued)

**Main Idea**

**Details**

### Clues to Earth's Interior

*I found this information on page _____ .*

**Complete** *the graphic organizer to explain how scientists use indirect observations to learn about Earth's interior.*

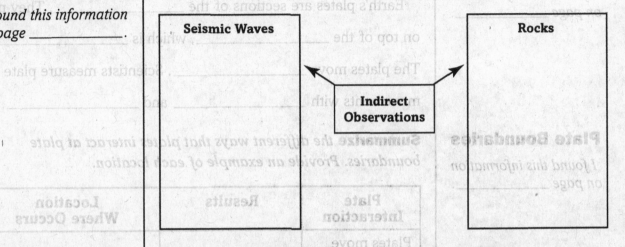

### Earth's Layers

*I found this information on page _____ .*

**Organize** *information about Earth's layers. Complete the outline.*

**Earth's Layers**

**A.** Inner core

  1. _____

  2. _____

  3. _____

**B.** Outer core

  1. _____

  2. _____

**C.** Mantle

  1. _____

  2. _____

**D.** Crust

  1. _____

  2. _____

  3. _____

Copyright © Glencoe/McGraw-Hill, a division of The McGraw-Hill Companies, Inc.

## Section 1 Earth's Moving Plates (continued)

**Main Idea**          **Details**

**Earth's Plates**

*I found this information on page _____.*

**Analyze** *Earth's plates. Fill in the missing words.*

Earth's plates are sections of the _____. They move

on top of the _____, which is _____.

The plates move _____. Scientists measure plate

movements with _____ and _____.

**Plate Boundaries**

*I found this information on page _____.*

**Summarize** *the different ways that plates interact at plate boundaries. Provide an example of each location.*

| Plate Interaction | Results | Location Where Occurs |
|---|---|---|
| Plates move apart. | | |
| Continental plates collide. | | |
| One plate sinks beneath another plate. | | |
| Plates slide past one another. | | |

---

**CONNECT IT**

Compare Earth's plates to a jigsaw puzzle. How are they similar?

_____

_____

_____

Copyright © Glencoe/McGraw-Hill, a division of The McGraw-Hill Companies, Inc.

# Forces Shaping Earth

**Section 2  Uplift of Earth's Crust**

**Scan** *Section 2. Then write three questions that occur to you.*

1. _____

2. _____

3. _____

**Review Vocabulary**  **Define** erosion *using your book or a dictionary.*

*erosion*  _____

**New Vocabulary**  *Write a sentence that reflects the scientific meaning of each vocabulary term.*

*fault-block mountain*  _____
_____

*folded mountain*  _____
_____

*upwarped mountain*  _____
_____

*volcanic mountain*  _____
_____

*isostasy*  _____
_____

**Academic Vocabulary**  *Write a two-line poem using the term* erode.

*erode*  _____
_____

Copyright © Glencoe/McGraw-Hill, a division of The McGraw-Hill Companies, Inc.

## Section 2  Uplift of Earth's Crust (continued)

**Main Idea**                                    **Details**

**Building
Mountains**

*I found this information
on page* _____.

**Identify** *the 4 main types of mountains.*

1. _____    3. _____

2. _____    4. _____

**Contrast** *mountains that are still forming with older mountains.*

Mountains that are still forming are _____ and _____.

Older mountains have _____.

*I found this information
on page* _____.

**Organize** *information from your book about fault-block, folded,
and upwarped mountains.*

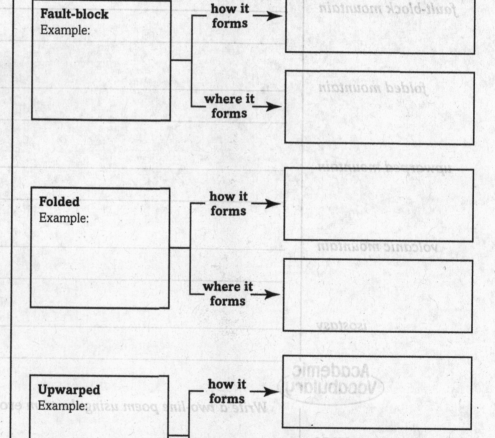

| Fault-block Example: | → how it forms → | |
| | → where it forms → | |

| Folded Example: | → how it forms → | |
| | → where it forms → | |

| Upwarped Example: | → how it forms → | |
| | → where it forms → | interior of continent |

Copyright © Glencoe/McGraw-Hill, a division of The McGraw-Hill Companies, Inc.

## Section 2  Uplift of Earth's Crust (continued)

**Main Idea**                    **Details**

### Building Mountains

*I found this information on page _____.*

**Create** *a cross-section drawing of a volcanic mountain formed on land. Show the magma, magma chamber, pipe, vent, and crater as the magma flows from underground out of the crater.*

**Compare and contrast** *volcanic mountains formed at subduction zones with those formed over hot spots. Complete the Venn diagram with at least 6 points of information from your book.*

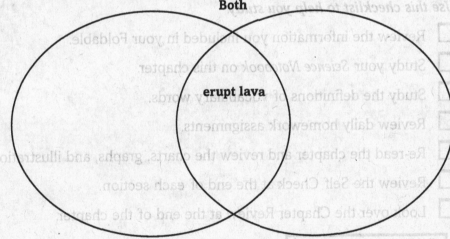

Subduction zone volcanoes        Hot spot volcanoes

Both

erupt lava

**CONNECT IT** Use what you have learned about isostasy to compare the crust under the Appalachian Mountains today with the crust when the mountains formed.

_____

_____

_____

Copyright © Glencoe/McGraw-Hill, a division of The McGraw-Hill Companies, Inc.

# Forces Shaping Earth Chapter Wrap-Up

*Review the ideas you listed in the K-W-L table at the beginning of the chapter. Cross out any incorrect information in the first column. Then complete the table by filling in the third column. How do your ideas now compare with those you provided at the beginning of the chapter?*

| K<br>What I know | W<br>What I want to find out | L<br>What I learned |
|---|---|---|
|  |  |  |
|  |  |  |
|  |  |  |

# Review

*Use this checklist to help you study.*

☐ Review the information you included in your Foldable.

☐ Study your *Science Notebook* on this chapter.

☐ Study the definitions of vocabulary words.

☐ Review daily homework assignments.

☐ Re-read the chapter and review the charts, graphs, and illustrations.

☐ Review the Self Check at the end of each section.

☐ Look over the Chapter Review at the end of the chapter.

## SUMMARIZE IT

After reading this chapter, identify three things that you have learned about forces that shape Earth.

_____

_____

_____

_____

Copyright © Glencoe/McGraw-Hill, a division of The McGraw-Hill Companies, Inc.

# Weathering and Erosion

## Before You Read

*Before you read the chapter, respond to these statements.*

1. Write an **A** if you agree with the statement.
2. Write a **D** if you disagree with the statement.

| Before You Read | Weathering and Erosion |
|---|---|
| | • Weathering is the conditions of the atmosphere at a given time. |
| | • Soil forms from pieces of broken rock and other kinds of matter. |
| | • Erosion moves rock and soil from one place to another. |
| | • Water can cause erosion, but ice cannot. |

**FOLDABLES™ Study Organizer**

*Construct the Foldable as directed at the beginning of this chapter.*

**Science Journal**

*Describe a place—a home, a park, a river, or a mountain. What might that place look like in a year, a hundred years, even 5,000 years?*

_____

_____

_____

_____

_____

_____

Copyright © Glencoe/McGraw-Hill, a division of The McGraw-Hill Companies, Inc.

# Weathering and Erosion
## Section 1  Weathering and Soil Formation

**Skim** through Section 1 of your book. Read the headings and look at the illustrations. Write three questions that come to mind.

1. _____

2. _____

3. _____

**Review Vocabulary**

**Define** the key terms using your book or a dictionary.

acid rain

**New Vocabulary**

weathering _____

mechanical weathering _____

chemical weathering _____

soil _____

topography _____

**Academic Vocabulary**

**Define** chemical *as an adjective. Use a dictionary to help you.*

chemical _____

Copyright © Glencoe/McGraw-Hill, a division of The McGraw-Hill Companies, Inc.

## Section 1 Weathering and Soil Formation (continued)

⬭ **Main Idea** ⬭                    ⬭ **Details** ⬭

**Weathering**

*I found this information on page* _____.

**Organize** *information by listing three things that cause rocks to weather.*

| Causes of Weathering | |
|---|---|
| 1. | |
| 2. | |
| 3. | |

**Mechanical Weathering**

*I found this information on page* _____.

**Identify** *major causes of mechanical weathering. Complete the concept map below.*

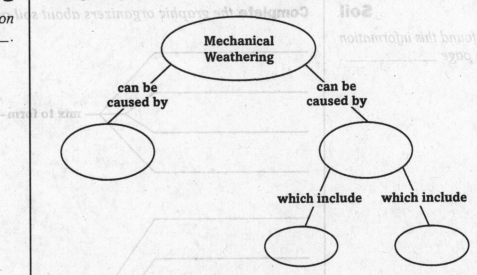

*I found this information on page* _____.

**Create** *three drawings to show the process of ice wedging.*

| | | |
|---|---|---|
| Water seeps into cracks. | Water freezes and expands, making cracks wider. | Ice melts and the process repeats. |

Copyright © Glencoe/McGraw-Hill, a division of The McGraw-Hill Companies, Inc.

## Section 1  Weathering and Soil Formation (continued)

**Main Idea**                                    **Details**

**Chemical Weathering**

*I found this information on page _____.*

**Organize** *the information from your book in the outline below.*

Chemical weathering

**A.** Definition: _____

_____

**B.** Causes:

1. _____

2. _____

3. _____

**Soil**

*I found this information on page _____.*

**Complete** *the graphic organizers about soil and soil formation.*

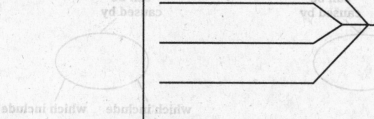

mix to form _____ soil

affect _____ soil formation

**CONNECT IT**  The temperature on some mountains is below freezing all year. Predict what soil on these mountains is like.

_____

_____

_____

Copyright © Glencoe/McGraw-Hill, a division of The McGraw-Hill Companies, Inc.

# Weathering and Erosion
## Section 2  Erosion of Earth's Surface

**Scan** *Use the checklist below to preview Section 2 of your book.*
*Then write three facts that you discovered about how erosion affects*
*Earth's surface.*

☐ Read all headings.

☐ Read all boldface words.

☐ Look at all of the pictures.

☐ Think about what you already know about features of
   Earth's surface.

1. _____

2. _____

3. _____

**Review Vocabulary**  *Write the correct vocabulary word next to each definition.*

_____    the dropping of sediment that occurs when an agent of erosion can
                    no longer carry its load

**New Vocabulary**

_____    the movement of rock or soil by gravity, ice, wind, or water

_____    erosion that occurs when gravity alone causes rock or sediment
                    to move down a slope

_____    the process in which sediment moves slowly downhill

_____    the movement of rock or sediment downhill along a curved surface

_____    the erosion of the land by wind

_____    erosion that occurs when wind blows sediment into rocks, makes
                    pits in the rocks, and produces a smooth, polished surface

_____    water that flows over the ground

**Academic Vocabulary**  **Define** occur *using a dictionary.*

*occur*    _____

Copyright © Glencoe/McGraw-Hill, a division of The McGraw-Hill Companies, Inc.

## Section 2 Erosion of Earth's Surface (continued)

Main Idea ———————— Details

**Agents of Erosion**

I found this information on page _____.

**Organize** *information from your book by filling in the concept map with the four agents, or causes, of erosion.*

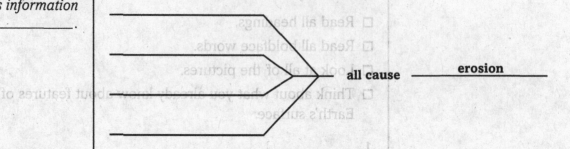

all cause          erosion

**Gravity**

I found this information on page _____.

**Compare and contrast** *the four types of mass movements. Write ways they are all the same and some ways they are different.*

| Mass Movements | |
|---|---|
| **Similarities** | **Differences** |
| | |
| | |
| | |
| | |

**Ice**

I found this information on page _____.

**Sequence** *four steps explaining how glaciers form and change Earth's surface.*

| Glaciers Form and Change Earth's Surface |
|---|
| 1. |
| 2. |
| 3. |
| 4. |

Copyright © Glencoe/McGraw-Hill, a division of The McGraw-Hill Companies, Inc.

Section 2  Erosion of Earth's Surface (continued)

## Main Idea

## Details

### Wind

*I found this information on page _____ .*

**Model** *how a sand dune moves by making a diagram in the box.*
*Label the following features:*

- sand blows up this side
- sand falls down this side
- dune movement (arrow)
- wind (arrow)

| | After You Read |
|---|---|
| Weathering is the conditions of the atmosphere at a given time. | |
| Soil forms from pieces of broken rock and other kinds of matter. | |
| Erosion moves rock and soil from one place to another. | |
| Water can cause erosion, but ice cannot. | |

### Water

*I found this information on page _____ .*

**Complete** *the concept map by listing several ways that water can flow over Earth's surface.*

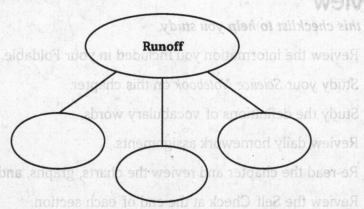

### Effects of Erosion

*I found this information on page _____ .*

**Analyze** *the effects of erosion. List three examples of landforms caused by erosion and three examples caused by deposition.*

| Effects of Erosion | |
|---|---|
| Where Sediment is Removed (erosion) | Where Sediment Accumulates (deposition) |
| | |
| | |

Copyright © Glencoe/McGraw-Hill, a division of The McGraw-Hill Companies, Inc.

# Weathering and Erosion
## Chapter Wrap-Up

*Now that you have read the chapter, think about what you have learned and complete the table below. Compare your previous answers with these.*

1. Write an **A** if you agree with the statement.
2. Write a **D** if you disagree with the statement.

| Weathering and Erosion | After You Read |
|---|---|
| • Weathering is the conditions of the atmosphere at a given time. | |
| • Soil forms from pieces of broken rock and other kinds of matter. | |
| • Erosion moves rock and soil from one one place to another. | |
| • Water can cause erosion, but ice cannot. | |

## Review

*Use this checklist to help you study.*

☐ Review the information you included in your Foldable.

☐ Study your *Science Notebook* on this chapter.

☐ Study the definitions of vocabulary words.

☐ Review daily homework assignments.

☐ Re-read the chapter and review the charts, graphs, and illustrations.

☐ Review the Self Check at the end of each section.

☐ Look over the Chapter Review at the end of the chapter.

**SUMMARIZE IT** After reading this chapter, identify three things that you have learned about weathering and erosion.

_____

_____

_____

_____

Copyright © Glencoe/McGraw-Hill, a division of The McGraw-Hill Companies, Inc.

# The Atmosphere in Motion

## Before You Read

*Before you read the chapter, respond to these statements.*

1. Write an **A** if you agree with the statement.
2. Write a **D** if you disagree with the statement.

| Before You Read | The Atmosphere in Motion |
|---|---|
| | • The atmosphere protects living things from harmful doses of ultraviolet radiation and X-ray radiation. |
| | • Earth is often referred to as the water planet. |
| | • Fast-moving molecules transfer energy to slower-moving molecules when they bump into them. |
| | • The highest layer of the atmosphere is the stratosphere. |

**FOLDABLES**
**Study Organizer**

*Construct the Foldable as directed at the beginning of this chapter.*

**Science Journal**

*Write a short newspaper article to warn people about the dangers of an approaching hurricane.*

_____

_____

_____

_____

_____

_____

Copyright © Glencoe/McGraw-Hill, a division of The McGraw-Hill Companies, Inc.

# The Atmosphere in Motion

## Section 1 The Atmosphere

**Scan** *Section 1 of your book. Use the checklist below.*

☐ Read all section titles.

☐ Read all boldface words.

☐ Read all charts and graphs.

☐ Look at all of the pictures.

☐ Think about what you already know about the atmosphere.

*Write three facts you discovered about the atmosphere as you scanned this section.*

1. _____

2. _____

3. _____

**Review Vocabulary**  **Define** evaporation *to show its scientific meaning.*

evaporation _____

_____

**New Vocabulary**  *Use your book to define the following terms.*

atmosphere _____

_____

aerosols _____

_____

water cycle _____

_____

**Academic Vocabulary**  *Use a dictionary to define* affect *as a verb.*

affect _____

Copyright © Glencoe/McGraw-Hill, a division of The McGraw-Hill Companies, Inc.

## Section 1  The Atmosphere (continued)

| ⬭ Main Idea ⬭ | ⬭ Details ⬭ |

### Investigating Air and **Composition of the Atmosphere**

*I found this information on page* _____.

**Complete** *the graphic organizer below to identify the ways that the atmosphere makes Earth fit for life.*

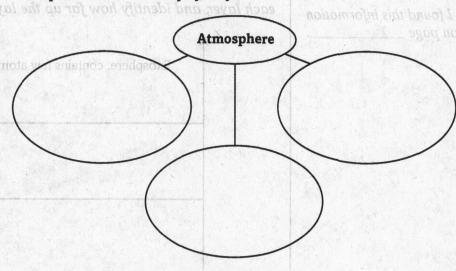

*I found this information on page* _____.

**Label** *the gases that form the three main components of the atmosphere, and indicate the percentage of each.*

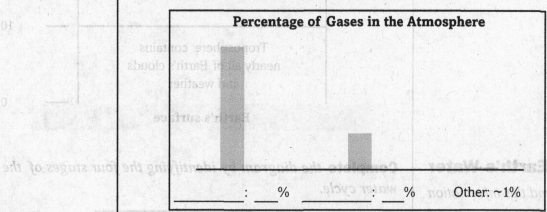

**Percentage of Gases in the Atmosphere**

_____ : ___ %     _____ : ___ %     Other: ~1%

*I found this information on page* _____.

**Summarize** *information about aerosols by completing the outline.*

**I.** Examples of aerosols

   **A.** Solids

     **1.** _____

     **2.** _____

     **3.** _____

   **B.** Tiny liquid droplets

     **1.** _____

Copyright © Glencoe/McGraw-Hill, a division of The McGraw-Hill Companies, Inc.

## Section 1  The Atmosphere (continued)

| **Main Idea** | **Details** |
|---|---|

**Layers of the Atmosphere**

I found this information on page _____.

**Organize** *information about the layers of the atmosphere by completing the diagram. Name and describe a characteristic of each layer, and identify how far up the layer extends.*

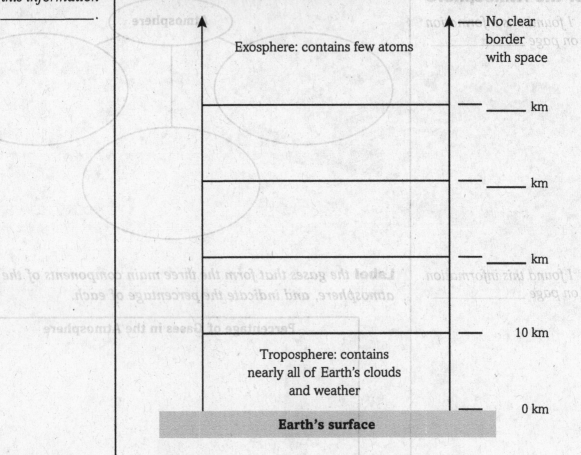

Exosphere: contains few atoms

————  No clear border with space

————  _____ km

————  _____ km

————  _____ km

————  10 km

Troposphere: contains nearly all of Earth's clouds and weather

————  0 km

**Earth's surface**

**Earth's Water**

I found this information on page _____.

**Complete** *the diagram by identifying the four stages of the water cycle.*

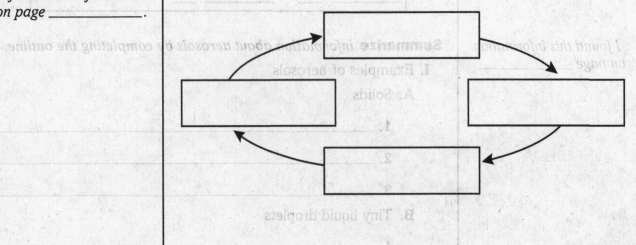

Copyright © Glencoe/McGraw-Hill, a division of The McGraw-Hill Companies, Inc.

# The Atmosphere in Motion
## Section 2  Earth's Weather

**Scan** *Section 2 of your book. Read the headings and look at the illustrations. Write three questions that come to mind.*

1. _____

2. _____

3. _____

**Review Vocabulary**  **Define** condensation *to show its scientific meaning.*

condensation _____

_____

**New Vocabulary**  *Use your book to define the following terms. Then write a sentence using each term.*

humidity _____

_____

_____

dew point _____

_____

_____

relative humidity _____

_____

_____

**Academic Vocabulary**  *Use a dictionary to define* factor *as a noun.*

factor _____

_____

Copyright © Glencoe/McGraw-Hill, a division of The McGraw-Hill Companies, Inc.

Section 2 **Earth's Weather** (continued)

| ⟨ **Main Idea** ⟩ | ⟨ **Details** ⟩ |

**Weather**

*I found this information
on page* _____ .

**Create** *a graphic organizer to identify the six weather factors.*

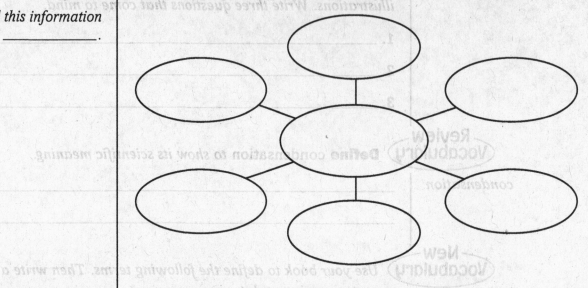

*I found this information
on page* _____ .

**Sequence** *how energy moves through the atmosphere by
completing the labels on the diagram.*

3. Cool air pushes warm air
upward, creating a

_____ .

2. Air at the surface
is heated by

_____ .

**Earth's surface**

1. Earth's surface is warmed by _____ .

Copyright © Glencoe/McGraw-Hill, a division of The McGraw-Hill Companies, Inc.

## Section 2 Earth's Weather (continued)

<table>
<tr><td colspan="2" align="center">**Main Idea**</td><td colspan="2" align="center">**Details**</td></tr>
<tr><td colspan="2">**Clouds**</td><td colspan="2">**Summarize** *types of clouds in the chart below.*</td></tr>
</table>

### Main Idea

**Clouds**

I found this information on page _____.

### Details

**Summarize** *types of clouds in the chart below.*

| Class | Altitude | Examples | |
|-------|----------|----------|--|
| Low | 2,000 m or below | cumulus, | a type that can extend from low to high: |
| Middle | | | |
| High | | | |

**Precipitation**

I found this information on page _____.

**Identify** *the different types of precipitation.*

1. _____  2. _____  3. _____

4. _____  5. _____

**Wind**

I found this information on page _____.

**Complete** *the diagram of Earth by identifying the major wind belts and drawing arrows to indicate the prevailing direction of the winds within each belt.*

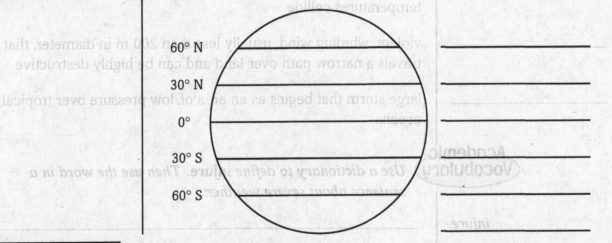

60° N _____

30° N _____

0° _____

30° S _____

60° S _____

---

**CONNECT IT** Explain how conduction warms bare feet when a person walks on hot sand along a beach.

_____

_____

_____

Copyright © Glencoe/McGraw-Hill, a division of The McGraw-Hill Companies, Inc.

# The Atmosphere in Motion

## Section 3  Air Masses and Fronts

**Predict** *three things that will be discussed as you read the headings of Section 3 of your book.*

1. _____

2. _____

3. _____

**Review Vocabulary**    **Define** thunderstorm *to show its scientific meaning.*

thunderstorm    _____

_____

**New Vocabulary**    *Write the terms to the left of their definitions.*

_____    large body of air that develops over a particular region of Earth's surface

_____    boundary that develops where air masses of different temperatures collide

_____    violent, whirling wind, usually less than 200 m in diameter, that travels a narrow path over land and can be highly destructive

_____    large storm that begins as an area of low pressure over tropical oceans

**Academic Vocabulary**    *Use a dictionary to define* injure. *Then use the word in a sentence about severe weather.*

injure    _____

Copyright © Glencoe/McGraw-Hill, a division of The McGraw-Hill Companies, Inc.

## Section 3  Air Masses and Fronts (continued)

<svg><!-- Main Idea --></svg> **Main Idea**          <svg><!-- Details --></svg> **Details**

### Air Masses

*I found this information on page _____.*

**Complete** *the blanks in the sentences about air masses.*

Air masses that _____ in one area for a few days pick up the _____ of that area. For example, an air mass that stays over a tropical ocean will become _____ and _____ .

### Fronts

*I found this information on page _____.*

**Contrast** *the four types of fronts by completing the chart.*

| Type of Front | How It Forms |
|---|---|
| Cold front | |
| | Warm air advances into region of colder air; the warm, less dense air slides up and over the colder air. |
| Stationary front | |
| | Fast-moving cold front overtakes a slower warm front. |

### High- and Low-Pressure Centers

*I found this information on page _____.*

**Compare** *ways that high pressure and low pressure affect weather.*

| High pressure forms. | → | Air _____ . | → | Moisture in air cannot _____ . | → | |

| Low pressure forms. | → | Air flows in and _____ . | → | Moisture in air _____ . | → | |

Copyright © Glencoe/McGraw-Hill, a division of The McGraw-Hill Companies, Inc.

## Section 3 Air Masses and Fronts (continued)

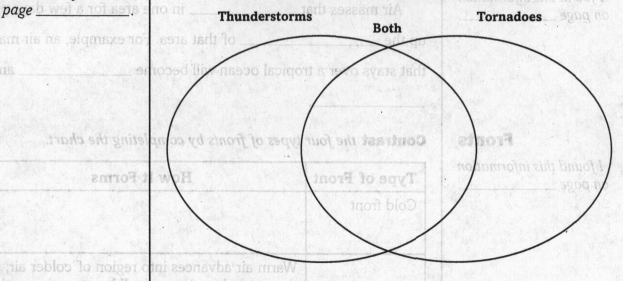

**Main Idea**

**Severe Weather**

*I found this information on page* _____.

**Details**

**Compare and contrast** *at least eight main characteristics of thunderstorms and tornadoes in the Venn diagram below.*

Thunderstorms       Both       Tornadoes

*I found this information on page* _____.

**Describe** *each of the following characteristics of a hurricane.*

**1.** Wind gusts _____

_____

**2.** Storm surge _____

_____

**3.** Beach erosion _____

_____

┌─ **CONNECT IT** ─┐

Explain the difference between a severe weather watch and a severe weather warning in terms of how you should respond to each.

_____

_____

_____

Copyright © Glencoe/McGraw-Hill, a division of The McGraw-Hill Companies, Inc.

# Tie It Together

## Model Sunlight on Earth

*Design a way to demonstrate how the curved surface of Earth can affect how much sunlight the equator receives versus how much the North Pole receives. Test your model. Write a list of detailed observations.*

_____

_____

_____

_____

_____

_____

_____

_____

_____

_____

_____

_____

_____

_____

_____

_____

_____

Copyright © Glencoe/McGraw-Hill, a division of The McGraw-Hill Companies, Inc.

# The Atmosphere in Motion
## Chapter Wrap-Up

*Now that you have read the chapter, think about what you have learned and complete the table below. Compare your previous answers with these.*

1. Write an **A** if you agree with the statement.
2. Write a **D** if you disagree with the statement.

| The Atmosphere in Motion | After You Read |
|---|---|
| • The atmosphere protects living things from harmful doses of ultraviolet radiation and X-ray radiation. | |
| • Earth is often referred to as the water planet. | |
| • Fast-moving molecules transfer energy to slower-moving molecules when they bump into them. | |
| • The highest layer of the atmosphere is the stratosphere. | |

# Review

*Use this checklist to help you study.*

☐ Review the information you included in your Foldable.

☐ Study your *Science Notebook* on this chapter.

☐ Study the definitions of vocabulary words.

☐ Review daily homework assignments.

☐ Re-read the chapter and review the charts, graphs, and illustrations.

☐ Review the Self Check at the end of each section.

☐ Look over the Chapter Review at the end of the chapter.

**SUMMARIZE IT** After reading this chapter, identify three things that you have learned about Earth's atmosphere.

_____

_____

_____

_____

Copyright © Glencoe/McGraw-Hill, a division of The McGraw-Hill Companies, Inc.

# Oceans

## Before You Read

*Before you read the chapter, respond to these statements.*

1. Write an **A** if you agree with the statement.
2. Write a **D** if you disagree with the statement.

| Before You Read | Oceans |
|---|---|
| | • Ocean water is about the same temperature all over the world. |
| | • Global winds cause density currents to move the ocean water. |
| | • The Moon's gravity affects the tides. |
| | • Wave erosion affects marine life in coastal regions. |

*Construct the Foldable as directed at the beginning of this chapter.*

**Science Journal**

*Write three questions that you would ask a scientist studying ocean life.*

_____

_____

_____

_____

_____

_____

_____

Copyright © Glencoe/McGraw-Hill, a division of The McGraw-Hill Companies, Inc.

Name _____  Date _____

# Oceans
## Section 1  Ocean Water

**Skim** *through Section 1 of your book. Write three questions that come to mind from reading the headings and the illustration captions.*

1. _____

2. _____

3. _____

**Review Vocabulary**

**Define** atmosphere *to show its scientific meaning.*

atmosphere

_____

_____

_____

**New Vocabulary**

*Define the following terms.*

salinity

_____

_____

photosynthesis

_____

thermocline

_____

**Academic Vocabulary**

*Define* accumulate. *Use* accumulate *in a sentence to show its scientific meaning.*

accumulate

_____

_____

_____

Copyright © Glencoe/McGraw-Hill, a division of The McGraw-Hill Companies, Inc.

Section 1 Ocean Water (continued)

## Main Idea

**Importance of Oceans**

*I found this information on page _____.*

**Formation of Oceans**

*I found this information on page _____.*

**Composition of Ocean Water**

*I found this information on page _____.*

## Details

**Identify** *four reasons the oceans are important as discussed in your book.*

1. _____
2. _____
3. _____
4. _____

**Sequence** *five steps to the formation of oceans, and write them in the correct order.*

| | |
|---|---|
| 1. | Gases and water vapor entered Earth's atmosphere. |
| 2. | |
| 3. | |
| 4. | |
| 5. | |

**Analyze** *the information in your book to complete the graphic organizer below.*

**Salinity**

Water makes up _____ percent of seawater. Of the remaining solids, the two most abundant elements in seawater are:

30.6%

55%

join to make

when water evaporates

Copyright © Glencoe/McGraw-Hill, a division of The McGraw-Hill Companies, Inc.

## Section 1  Ocean Water (continued)

### Main Idea

**Composition of Ocean Water** and **Water Temperature and Pressure**

*I found this information on page _____.*

### Details

**Outline** *the material about dissolved gases and ocean temperature.*

I. The three most important gases are

_____, _____, _____

A. Oxygen gas

1. _____

_____

2. _____

B. Carbon dioxide gas

1. _____

2. _____

3. _____

C. Nitrogen gas

1. _____

2. _____

_____

a. _____

b. _____

II. Oceans have three temperature layers.

A. _____

1. _____

2. _____

B. Thermocline layer

C. _____

**CONNECT IT**  Scuba divers don't need the pressurized suits that deep-sea divers do. Hypothesize why deep-sea divers must use special equipment. _____

_____

Copyright © Glencoe/McGraw-Hill, a division of The McGraw-Hill Companies, Inc.

# Oceans

## Section 2  Ocean Currents and Climate

**Scan** *the list below to preview Section 2 of your book.*

- Read all section titles.
- Read all bold words.
- Read all charts and graphs
- Look at all the pictures and read their captions.
- Think about what you already know about oceans.

*Write four facts you discovered about oceans as you scanned the section.*

1. _____
2. _____
3. _____
4. _____

**Review Vocabulary**   **Define** current *to show its scientific meaning.*

current   _____

**New Vocabulary**   *Define the following terms.*

surface current   _____
_____

density current   _____
_____

upwelling   _____
_____

**Academic Vocabulary**   **Define** distribute *to show its scientific meaning.*

distribute   _____

## Section 2 Ocean Currents and Climate (continued)

<table>
<tr><td>

**Main Idea**

</td><td>

**Details**

</td></tr>
<tr><td>

**Surface Currents**

*I found this information on page* _____.

</td><td>

**Complete** *the following sentences on surface currents.*

When _____ blow across the ocean's surface, they can set ocean water in motion. Ocean currents do not move in straight lines because of _____. In the northern hemisphere, currents circulate clockwise, or to the _____. In the southern hemisphere, currents circulate counter-clockwise, or to the _____. This is an example of the _____.

</td></tr>
<tr><td>

*I found this information on page* _____.

</td><td>

**The Gulf Stream**

**Label** *Use the surface current map in your book and the directions below to trace the Gulf Steam and other surface currents.*

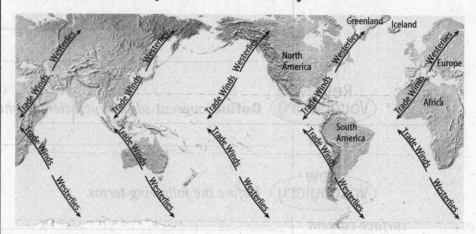

• Trace the equator in black.

• The Gulf Stream flows from Florida, northeast toward North Carolina, and then across the Atlantic Ocean. Trace and label the Gulf Stream in green.

• Currents that originate near the equator are warm. Trace these currents in red.

• Use blue to trace the currents on the western coasts of continents that return cold water back toward the equator.

• The warm Gulf Stream keeps Iceland's climate mild and its harbors ice-free year-round. Outline Iceland in orange.

</td></tr>
</table>

Copyright © Glencoe/McGraw-Hill, a division of The McGraw-Hill Companies, Inc.

## Section 2 Ocean Currents and Climate (continued)

Main Idea　Details

**Density Currents**

*I found this information on page _____.*

**Skim** *the information on density currents. In the Question spaces below, turn the bold-faced text headings into questions. The first one has been done for you. Then answer your questions.*

**Density Currents**

Question: *How are density currents formed?* _____

Answer: _____

_____

**Cold and Salty Water**

Question: _____?

Answer: _____

_____

_____

_____

**Density Currents and Climate Change**

Question: _____?

Answer: _____

_____

**Upwelling**

*I found this information on page _____.*

**Sequence** *the steps in the process of upwelling.*

| 1. | Winds cause surface water to move away from the land because of the Coriolis effect. |
|----|-------------------------------------------------------------------------------------|
| 2. | |
| 3. | |
| 4. | |

Copyright © Glencoe/McGraw-Hill, a division of The McGraw-Hill Companies, Inc.

# Oceans
## Section 3  Waves

**Predict** *Read the title of Section 3. List three things that might be discussed in this section.*

1. _____

2. _____

3. _____

**Review Vocabulary**

*Use* sediments *in a scientific sentence.*

sediments _____

_____

**New Vocabulary**

*Locate and write the sentence where the new word appears.*

wave _____

_____

tide _____

_____

**Academic Vocabulary**

**Define** range *to show its scientific meaning.*

range _____

_____

Copyright © Glencoe/McGraw-Hill, a division of The McGraw-Hill Companies, Inc.

### Section 3 Waves (continued)

| Main Idea | Details |

**Waves Caused by Wind**

*I found this information on page _____.*

**Sequence** *the process of wave formation.*

1. **Wind blows across a body of water.**

2.

3.

4. **A wave forms.**

**Organize** *the 3 factors that affect the height of a wave.*

_____

_____

_____

*I found this information on page _____.*

**Identify** *the parts of a wave using the terms below.*

| breaker | trough | crest |
| wave height | swells | wave length |

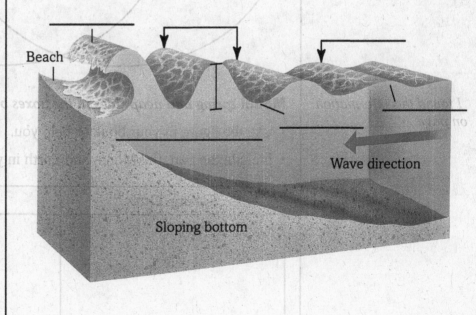

Beach

Wave direction

Sloping bottom

Copyright © Glencoe/McGraw-Hill, a division of The McGraw-Hill Companies, Inc.

Section 3 **Waves** (continued)

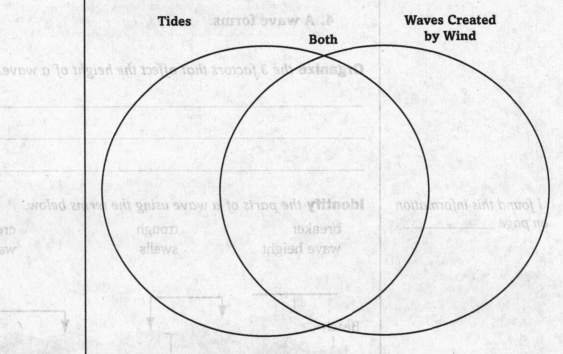

| Main Idea | Details |
|---|---|

**Tides**

*I found this information on page _____.*

**Organize** *each wave characteristic in the Venn diagram to show whether it is a trait of tides, waves created by wind, or both.*

- breakers
- higher different times of the month
- contain crests and troughs
- wavelength can be measured

- created by the Sun's and the Moon's gravity
- created by winds
- wave height can be measured

**Tides**     **Both**     **Waves Created by Wind**

*I found this information on page _____.*

**Model** *spring and neap tides in the boxes below.*

- Use the figure in your book to help you.
- Include the Sun, the Moon, and Earth in your model drawings.

| Spring Tide | Neap Tide |
|---|---|
|  |  |

Copyright © Glencoe/McGraw-Hill, a division of The McGraw-Hill Companies, Inc.

# Oceans
## Section 4 Life in the Oceans

**Create** *a postcard of the ocean that is interesting and educational. Use the photos and captions in Section 4 for ideas.*

Copyright © Glencoe/McGraw-Hill, a division of The McGraw-Hill Companies, Inc.

### Review Vocabulary

nutrients

*Use* nutrients *in a scientific sentence.*

_____

_____

### New Vocabulary

**Define** *Read the definitions below. Write the key term on the blank in the left column.*

_____  tiny marine organisms that drift in the surface waters of every ocean

_____  marine animals that actively swim in ocean waters—turtles and fish

_____  an organism that obtains food by eating other organisms

_____  organism that can make its own food by photosynthesis or chemosynthesis

_____  process in which bacteria make food from dissolved sulfur compounds

_____  organism that breaks down tissue and releases nutrients and carbon dioxide back into the ecosystem

## Section 4 Life in the Oceans (continued)

### ⟨Main Idea⟩                               ⟨Details⟩

**Types of Ocean Life**

*I found this information on page _____.*

**Classify** *plankton, nekton, or bottom dweller beside its description. Give three examples of each.*

| Description | Organism | Examples |
|---|---|---|
| burrow, walk, swim, can attach to the bottom | | 1.<br>2.<br>3. |
| actively swim in ocean | | 1.<br>2.<br>3. |
| usually one-celled marine organisms that float in ocean currents | | 1.<br>2.<br>3. |

**Ocean Ecosystems**

*I found this information on page _____.*

**Create** *an ocean ecosystem with four producers, four consumers, and two decomposers.*

- Label each organism as a producer, consumer, or decomposer.
- Draw arrows to show the transfer of energy in the food chain.

How many different transfer-of-energy arrows did you use to

connect the organisms in your food chain? _____

In all ecosystems, food chains are interconnected to form highly

complex systems called _____.

Copyright © Glencoe/McGraw-Hill, a division of The McGraw-Hill Companies, Inc.

## Section 4  Life in the Oceans (continued)

**Main Idea**                                    **Details**

### Ocean Nutrients

*I found this information on page _____.*

**Complete** *the chart below to describe how carbon is absorbed and released by the different parts of an ocean ecosystem.*

|  | Where does carbon come from? | Where does carbon go? |
|---|---|---|
| Atmosphere |  |  |
| Ocean Water |  |  |
| Producers |  |  |
| Consumers |  |  |
| Sediments |  |  |

**CONNECT IT**  Infer from your reading three ways that coral reefs are protected from pollution and habitat destruction. _____

_____

_____

_____

Copyright © Glencoe/McGraw-Hill, a division of The McGraw-Hill Companies, Inc.

# Oceans Chapter Wrap-Up

*Now that you have read the chapter, think about what you have learned and complete the table below. Compare your previous answers with these.*

1. Write an **A** if you agree with the statement.
2. Write a **D** if you disagree with the statement.

| Oceans | After You Read |
|---|---|
| • Ocean water is about the same temperature all over the world. | |
| • Global winds cause density currents to move the ocean water. | |
| • The Moon's gravity affects the tides. | |
| • Wave erosion affects marine life in coastal regions. | |

# Review

*Use this checklist to help you study.*

☐ Review the information you included in your Foldable.

☐ Study your *Science Notebook* on this chapter.

☐ Study the definitions of vocabulary words.

☐ Review daily homework assignments.

☐ Re-read the chapter and review the charts, graphs, and illustrations.

☐ Review the Self Check at the end of each section.

☐ Look over the Chapter Review at the end of the chapter.

**SUMMARIZE IT** After reading this chapter, identify three things that you have learned about oceans.

_____

_____

_____

_____

Copyright © Glencoe/McGraw-Hill, a division of The McGraw-Hill Companies, Inc.

Copyright © Glencoe/McGraw-Hill, a division of The McGraw-Hill Companies, Inc.

# Exploring Space

## Before You Read

*Preview the chapter, including section titles and the section headings. Complete the chart by listing at least one idea for each of the three sections in each column.*

| K<br>What I know | W<br>What I want to find out |
|---|---|
|  |  |
|  |  |
|  |  |
|  |  |

**FOLDABLES**
**Study Organizer**

*Construct the Foldable as directed at the beginning of this chapter.*

**Science Journal**

*Do you think space exploration is worth the risk and expense? Explain why.*

_____

_____

_____

_____

_____

_____

_____

# Exploring Space

## Section 1  Radiation from Space

**Evaluate** *the objectives found in* What You'll Learn *for Section 1. Write three questions that come to mind from reading these statements.*

1. _____
2. _____
3. _____

**Review Vocabulary**  **Define** telescope *using your book or a dictionary.*

telescope _____
_____

**New Vocabulary**  *Use your book or a dictionary to define the vocabulary terms.*

electromagnetic spectrum _____

refracting telescope _____

reflecting telescope _____
_____

observatory _____
_____

radio telescope _____
_____

**Academic Vocabulary**  *Use a dictionary to define* design *as a verb.*

design _____
_____

Copyright © Glencoe/McGraw-Hill, a division of The McGraw-Hill Companies, Inc.

Name _____ Date _____

## Section 1 Radiation from Space (continued)

```
  Main Idea                                    Details
```

### Electromagnetic Waves

*I found this information on page _____.*

**List** *seven forms of* electromagnetic radiation.

1. _____    5. _____
2. _____    6. _____
3. _____    7. _____
4. _____

**Compare and contrast** short wavelength radiation *with* long wavelength radiation *by completing the chart below.*

|  | Short Wavelength | Long Wavelength |
|---|---|---|
| Sketch of each wave |  |  |
| Description of frequency |  |  |

### Optical Telescopes

*I found this information on page _____.*

**Compare** *a* refracting telescope *with a* reflecting telescope.
- Use your book to help you draw cross-sections of each telescope.
- Use arrows to indicate the path taken by light in each type.
- Label the eyepiece lens, focal point, and any other mirrors or lenses.
- Model the shapes of a convex lens and a concave mirror.

**refracting telescope**          **reflecting telescope**

**convex lens**                   **concave mirror**

Copyright © Glencoe/McGraw-Hill, a division of The McGraw-Hill Companies, Inc.

Section 1 Radiation from Space (continued)

⟨ **Main Idea** ⟩ _____ ⟨ **Details** ⟩

### Optical Telescopes

I found this information on page _____.

**Summarize** *information about the* Hubble Space Telescope *by completing the paragraph.*

In _____, the _____

was launched. Scientists expected clear pictures from this

_____ telescope because it was _____

_____. However, a mistake was made when

the telescope's _____, so it did

not make _____. Repair missions were made in

(years) _____, when small _____

were added to correct the images.

### Radio Telescopes

I found this information on page _____.

**Organize** *information about* radio telescopes *in the chart below.*

| Radio telescopes |
|---|
| **Purpose:** |
| **Design:** |
| **Collect information used to:**<br><br>1.     3.<br><br>2. |

**CONNECT IT**
Radio waves from space have been studied for decades, but
scientists have yet to find signs of intelligent life. Suggest several explanations for this.

_____

_____

_____

_____

Copyright © Glencoe/McGraw-Hill, a division of The McGraw-Hill Companies, Inc.

# Exploring Space
## Section 2  Early Space Missions

**Predict** *three things that you think might be discussed in this section after reading its headings.*

1. _____

2. _____

3. _____

**Review Vocabulary**  *Write the correct vocabulary term next to each definition.*

_____  force that propels an aircraft or missile

**New Vocabulary**

_____  curved path followed by a satellite as it revolves around an object

_____  space mission with goal of landing a human on the Moon's surface

_____  special engine that can work in space and burns liquid or solid fuel

_____  space mission with goals of connecting spacecraft in orbit and investigating the effects of space travel on the human body

_____  any object that revolves around another object in space

_____  space mission with goal of orbiting a piloted spacecraft around Earth and bringing it back safely

_____  instrument that gathers information and sends it back to Earth

**Academic Vocabulary**  *Define the scientific meaning of* **goal** *using a dictionary.*

*goal* _____

_____

_____

Copyright © Glencoe/McGraw-Hill, a division of The McGraw-Hill Companies, Inc.

Section 2 **Early Space Missions** (continued)

## Main Idea

## Details

**The First Missions into Space**

*I found this information on page _____.*

**Compare and contrast** *the* two types of rockets *by completing the Venn diagram with the information below.*

- can be shut down and restarted
- do not require air for operation
- liquid fuel and oxidizer stored in separate tanks
- preferred for long-term space missions

- gases thrust it forward
- rubberlike fuel contains oxidizer
- generally simpler
- cannot be shut down once ignited

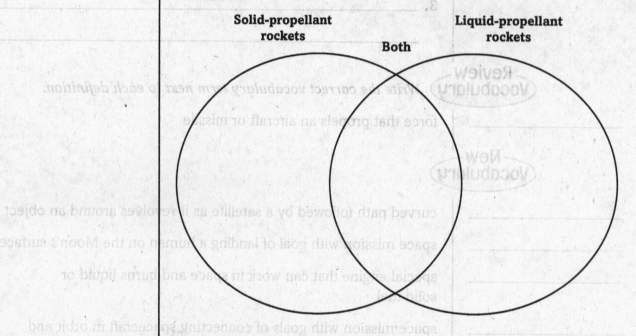

Solid-propellant rockets          **Both**          Liquid-propellant rockets

*I found this information on page _____.*

**Model** *the* path of a satellite. *Draw a satellite in orbit around Earth. Show the complete path of the satellite and the path it would take if it were not affected by gravity.*

Copyright © Glencoe/McGraw-Hill, a division of The McGraw-Hill Companies, Inc.

Section 2  Early Space Missions (continued)

~Main Idea~ _____ ~Details~ _____

### Space Probes

*I found this information on page _____.*

**Compare** *the advantages and disadvantages of* space probes *with* spacecraft piloted by humans.

| Comparison of Space Probes to Piloted Spacecraft | |
|---|---|
| Advantages | Disadvantages |
|  |  |

### Moon Quest

*I found this information on page _____.*

**Create** *a time line of the* United States' quest to reach the Moon *by identifying an event that corresponds to each date.*

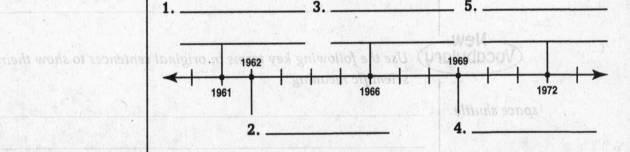

1. _____  3. _____  5. _____

1961  1962  1966  1969  1972

2. _____  4. _____

**CONNECT IT**  Design a plan for a space mission to take humans to Mars. Analyze challenges the crew would have to face. Develop a simple program to help prepare the crew to face these challenges.

_____

_____

_____

_____

_____

Copyright © Glencoe/McGraw-Hill, a division of The McGraw-Hill Companies, Inc.

# Exploring Space
## Section 3 Current and Future Space Missions

**Skim** *through Section 3 of your text. Read the headings and examine the illustrations. Write three questions that come to mind. Try to answer your questions as you read.*

1. _____

2. _____

   _____

3. _____

**Review Vocabulary**

*Use* cosmonaut *in a sentence that shows its scientific meaning.*

cosmonaut

_____

_____

**New Vocabulary**

*Use the following key terms in original sentences to show their scientific meaning.*

space shuttle _____

_____

space station _____

_____

**Academic Vocabulary**

**Define** *the scientific meaning of technology using a dictionary.*

technology _____

_____

_____

Copyright © Glencoe/McGraw-Hill, a division of The McGraw-Hill Companies, Inc.

## Section 3 Current and Future Space Missions (continued)

### Main Idea

### Details

**The Space Shuttle**

*I found this information
on page _____.*

**Summarize** *characteristics of the* space shuttle *below.*

| Engines: | Cargo bay: |
|---|---|
| **Landings:** | **Reusability:** |

**Exploring Mars**

*I found this information
on page _____.*

**Organize** *information about* missions to Mars *by completing the diagram. Identify each probe by its name and mission.*

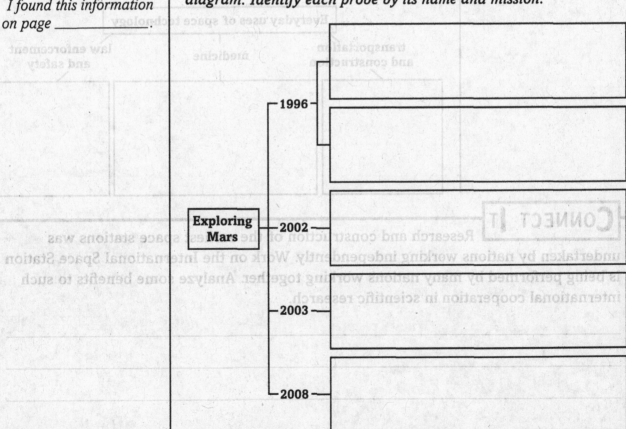

Copyright © Glencoe/McGraw-Hill, a division of The McGraw-Hill Companies, Inc.

## Section 3  Current and Future Space Missions (continued)

### ◁Main Idea▷                                    ◁Details▷

**Exploring the Moon** and *Cassini*

*I found this information on page _____.*

**Complete** *the chart with information about the* Lunar Prospector *and* Cassini *spacecraft.*

| Spacecraft | Launch Date | Destination | Goals |
|---|---|---|---|
| Lunar Prospector | | | |
| Cassini | | | |

*I found this information on page _____.*

**Organize** *information by identifying an example of technology developed for space programs that is useful in everyday life.*

```
            Everyday uses of space technology
   ┌──────────────────┬──────────────┬──────────────────┐
transportation      medicine      law enforcement
and construction                   and safety
   ┌──────┐          ┌──────┐          ┌──────┐
   │      │          │      │          │      │
   └──────┘          └──────┘          └──────┘
```

┌─ **CONNECT IT** ─────────────────────────────────────────┐

Research and construction of the earliest space stations was undertaken by nations working independently. Work on the International Space Station is being performed by many nations working together. Analyze some benefits to such international cooperation in scientific research.

_____

_____

_____

_____

Copyright © Glencoe/McGraw-Hill, a division of The McGraw-Hill Companies, Inc.

# Tie It Together

## Synthesize It

*Much of today's planetary research is carried out using remote-controlled rovers that are monitored and maneuvered by scientists on Earth. Suppose that you could design a remote-controlled rover to conduct research on a planet or the Moon.*

- Draw a sketch of your rover below.
- Identify features you would include on your rover.
- Explain why you would include each feature.
- Use what you have learned about space technologies in this section.

Copyright © Glencoe/McGraw-Hill, a division of The McGraw-Hill Companies, Inc.

# Exploring Space  Chapter Wrap-Up

*Review the ideas you listed in the chart at the beginning of the chapter. Cross out any incorrect information in the first column. Then complete the chart by filling in the third column.*

| K<br>What I know | W<br>What I want to find out | L<br>What I learned |
|---|---|---|
|  |  |  |
|  |  |  |

## Review

*Use this checklist to help you study.*

☐ Review the information you included in your Foldable.

☐ Study your *Science Notebook* on this chapter.

☐ Study the definitions of vocabulary words.

☐ Review daily homework assignments.

☐ Re-read the chapter and review the charts, graphs, and illustrations.

☐ Review the Self Check at the end of each section.

☐ Look over the Chapter Review at the end of the chapter.

### SUMMARIZE IT
After reading this chapter, identify three main ideas that you have learned about exploring space.

_____

_____

_____

_____

Copyright © Glencoe/McGraw-Hill, a division of The McGraw-Hill Companies, Inc.

# The Solar System and Beyond

## Before You Read

*Before you read the chapter, respond to these statements.*

1. Write an **A** if you agree with the statement.
2. Write a **D** if you disagree with the statement.

| Before You Read | The Solar System and Beyond |
|---|---|
| | • The Sun appears each day because of Earth's rotation. |
| | • Earth's tilted axis combined with Earth's revolution around the Sun produces the seasons. |
| | • The solar system includes only the nine planets that orbit the Sun. |
| | • Temperature differences cause stars to be different sizes. |

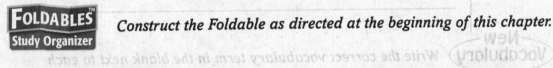

**FOLDABLES™**
**Study Organizer**

*Construct the Foldable as directed at the beginning of this chapter.*

**Science Journal**

*Write a short story about what it would be like to ride on a comet as it orbits the Sun.*

_____

_____

_____

_____

_____

_____

_____

_____

Copyright © Glencoe/McGraw-Hill, a division of The McGraw-Hill Companies, Inc.

# The Solar System and Beyond
## Section 1  Earth's Place in Space

**Skim** *Section 1 of your book. Read the headings. Write three questions that come to mind.*

1. _____

2. _____

3. _____

**Define** axis *using your book.*

axis  _____

_____

**New Vocabulary**  *Write the correct vocabulary term in the blank next to each definition.*

_____  the alternating rise and fall in sea level

_____  an event in which the Sun or Moon appears to grow dim due to a shadow cast by another body in space

_____  the spinning of Earth on its axis

_____  the movement of Earth around the Sun

_____  a regular curved path around the Sun

**Academic Vocabulary**  *Use a dictionary to define the word* visible. *Then use the word in a sentence about the solar system.*

visible  _____

_____

Copyright © Glencoe/McGraw-Hill, a division of The McGraw-Hill Companies, Inc.

## Section 1 Earth's Place in Space (continued)

**Main Idea** ———— **Details** ————

### Earth Moves

*I found this information on page _____.*

**Create** *a concept map about Earth's movement. Include information about Earth's rotation and revolution.*

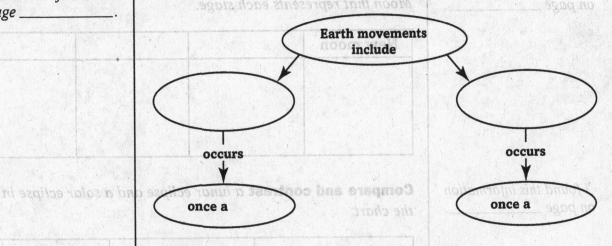

*I found this information on page _____.*

**Model** *why Earth experiences day and night.*

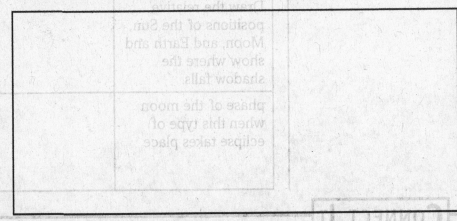

*I found this information on page _____.*

**Contrast** *in the chart below Earth's position compared to the Sun during summer and winter in the northern hemisphere.*

| Season | Earth's Tilt | Sun's Position in Sky |
|--------|--------------|----------------------|
| Summer |              |                      |
| Winter |              |                      |

Copyright © Glencoe/McGraw-Hill, a division of The McGraw-Hill Companies, Inc.

Section 1 Earth's Place in Space (continued)

**Main Idea** ───── **Details**

**Earth's Moon**

I found this information on page _____.

**Sequence** *the following stages of the lunar cycle: new moon, full moon, waning moon, waxing moon. **Then draw a picture of the Moon that represents each stage.***

| New moon | | | |
|---|---|---|---|
| | | | |

I found this information on page _____.

**Compare and contrast** *a lunar eclipse and a solar eclipse in the chart.*

| | Solar Eclipse | Lunar Eclipse |
|---|---|---|
| Draw the relative positions of the Sun, Moon, and Earth and show where the shadow falls. | | |
| phase of the moon when this type of eclipse takes place | | |

**CONNECT IT** Find out the date of the next new moon. Using what you have learned in this section about the Moon's phases, predict the date of the following first quarter moon, full moon, and third quarter moon. Sketch the phases of the moon.

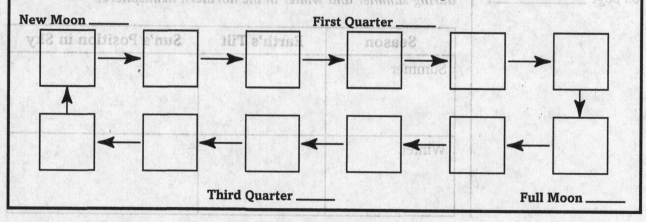

New Moon _____    First Quarter _____

Third Quarter _____    Full Moon _____

Copyright © Glencoe/McGraw-Hill, a division of The McGraw-Hill Companies, Inc.

Name _____ Date _____

# The Solar System and Beyond
## Section 2  The Solar System

**Scan** *Section 2 of your book. Write three facts that you discovered about the solar system as you scanned the section.*

1. _____

2. _____

3. _____

### Review Vocabulary
**Define** *system using your book or a dictionary.*

system _____

_____

### New Vocabulary
*Use your book or a dictionary to define the following terms.*

solar system _____

_____

astronomical unit _____

_____

comet _____

_____

meteorite _____

_____

### Academic Vocabulary
*Use a dictionary to define expose.*

expose _____

_____

Copyright © Glencoe/McGraw-Hill, a division of The McGraw-Hill Companies, Inc.

Section 2 The Solar System (continued)

## Main Idea · Details

### Distances in Space

*I found this information on page _____.*

**Evaluate** *why scientists decided to make an astronomical unit equal to the average distance between Earth and the Sun instead of choosing some other distance, such as that between Earth and the Moon.*

_____

_____

_____

_____

### Inner Planets

*I found this information on page _____.*

**Complete** *the outline below about the features of the inner planets.*

I. Mercury

   **A.** Atmosphere _____

   **B.** Temperature _____

   **C.** Surface _____

II. Venus

   **A.** Atmosphere _____

   **B.** Temperature _____

III. Earth

   **A.** Atmosphere _____

   **B.** Temperature _____

   _____

   **C.** Surface _____

IV. Mars

   **A.** Surface _____

   **B.** Water _____

   _____

Copyright © Glencoe/McGraw-Hill, a division of The McGraw-Hill Companies, Inc.

Section 2  The Solar System (continued)

## Main Idea

### Inner Planets and Outer Planets

*I found this information on page* _____.

## Details

**Compare** *the inner planets and outer planets by inserting the phrases into the Venn diagram.*

- closer to the Sun
- rocky planets
- some have rings
- orbit the Sun

- include gas giants
- include Earth
- farther from the Sun

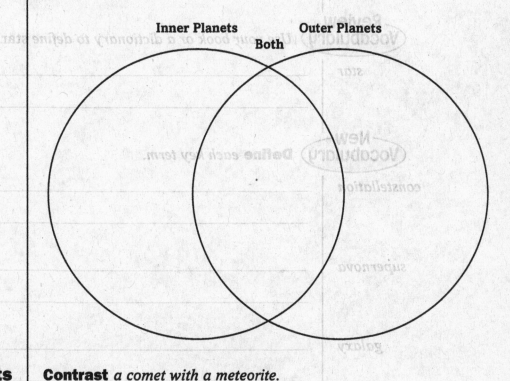

Inner Planets    Both    Outer Planets

### Comets

*I found this information on page* _____.

**Contrast** *a comet with a meteorite.*

_____

_____

_____

| **CONNECT IT** | Describe factors that make it impossible for humans to visit some |

planets in our solar system.

_____

_____

_____

Copyright © Glencoe/McGraw-Hill, a division of The McGraw-Hill Companies, Inc.

# The Solar System and Beyond
## Section 3  Stars and Galaxies

**Skim** *the headings in Section 3. Then make 3 predictions about what you will learn.*

1. _____

2. _____

3. _____

### Review Vocabulary

*Use your book or a dictionary to define* star.

star _____

_____

### New Vocabulary

**Define** *each key term.*

constellation _____

_____

_____

supernova _____

_____

galaxy _____

_____

light-year _____

_____

_____

### Academic Vocabulary

*Use a dictionary to define* apparent.

apparent _____

_____

Copyright © Glencoe/McGraw-Hill, a division of The McGraw-Hill Companies, Inc.

## Section 3 Stars and Galaxies (continued)

### Main Idea | Details

**Stars**

*I found this information on page _____.*

**Synthesize** *information from your book to write a short paragraph describing what you might see if you were to gaze at the stars for an entire night.*

_____
_____
_____
_____
_____
_____

**The Lives of Stars**

*I found this information on page _____.*

**Sequence** *the phrases to show the life cycle of a medium-sized star.*

- black dwarf forms
- cloud of dust and gas contracts
- main sequence star forms
- temperature rises at center of cloud

- star's core collapses
- fusion begins
- white dwarf forms
- star expands to become a giant

1. _____
2. _____
3. _____
4. _____
5. _____
6. _____
7. _____
8. _____

Copyright © Glencoe/McGraw-Hill, a division of The McGraw-Hill Companies, Inc.

## Section 3  Stars and Galaxies (continued)

**Main Idea**                                              **Details**

**Galaxies**

*I found this information on page _____.*

**Summarize** *the 3 types of galaxies.*

| Galaxies | |
|---|---|
| Category | Description |
| | |
| | |
| | |

*I found this information on page _____.*

**Model** *the Milky Way by making a sketch. Label the 5 arms and the location of our Sun. Use the figure in your book to help you.*

**The Universe**

*I found this information on page _____.*

**Complete** *the blanks in the paragraph below.*

Each _____ contains billions of stars. As many as

_____ galaxies might exist. All of these galaxies with

all of their billions of stars make up the _____.

**CONNECT IT**  The stars in the universe have been compared to the grains of sand on Earth. Write a sentence to explain this comparison.

_____

_____

Copyright © Glencoe/McGraw-Hill, a division of The McGraw-Hill Companies, Inc.

# Tie It Together

## Synthesize

*Imagine that you have just completed a trip through the universe. Write a journal entry or a story that you would tell your friends about your trip. Include what you saw and how you interacted with your surroundings.*

_____

_____

_____

_____

_____

_____

_____

_____

_____

_____

_____

_____

_____

_____

_____

_____

_____

_____

_____

_____

_____

Copyright © Glencoe/McGraw-Hill, a division of The McGraw-Hill Companies, Inc.

# The Solar System and Beyond
## Chapter Wrap-Up

*Now that you have read the chapter, think about what you have learned and complete the table below. Compare your previous answers with these.*

1. Write an **A** if you agree with the statement.
2. Write a **D** if you disagree with the statement.

| The Solar System and Beyond | After You Read |
|---|---|
| • The Sun appears each day because of Earth's rotation. | |
| • Earth's tilted axis combined with Earth's revolution around the Sun produces the seasons. | |
| • The solar system includes only the nine planets that orbit the Sun. | |
| • Temperature differences cause stars to be different sizes. | |

# Review

*Use this checklist to help you study.*

☐ Review the information you included in your Foldable.

☐ Study your *Science Notebook* on this chapter.

☐ Study the definitions of vocabulary words.

☐ Review daily homework assignments.

☐ Re-read the chapter and review the charts, graphs, and illustrations.

☐ Review the Self Check at the end of each section.

☐ Look over the Chapter Review at the end of the chapter.

## SUMMARIZE IT
After reading this chapter, identify three things that you have learned about the solar system.

_____

_____

_____

_____

Copyright © Glencoe/McGraw-Hill, a division of The McGraw-Hill Companies, Inc.

# Cells—The Units of Life

## Before You Read

*Before you read the chapter, respond to these statements.*

1. Write an **A** if you agree with the statement.

2. Write a **D** if you disagree with the statement.

| Before You Read | Cells—The Units of Life |
|---|---|
| | • Bacteria are the smallest organisms on Earth. |
| | • All living things are made up of one or more cells. |
| | • A cell's shape and size can be related to its function. |
| | • Cells are organized into systems to perform functions that keep an organism alive. |

**Construct the Foldable as directed at the beginning of this chapter.**

**Science Journal**

*Describe how building blocks fit together to build a larger structure.*

_____

_____

_____

_____

_____

_____

Copyright © Glencoe/McGraw-Hill, a division of The McGraw-Hill Companies, Inc.

# Cells—The Units of Life
## Section 1 The World of Cells

**Skim** through Section 1 of your text. Write three questions that come to mind.

1. _____

2. _____

3. _____

**Review Vocabulary**

Use the term **theory** in a sentence to illustrate its scientific meaning.

_theory_

_____

_____

**New Vocabulary**

Use the following key terms in a sentence to reflect their scientific meanings.

_bacteria_

_____

_____

_cell wall_

_____

_____

_organelle_

_____

_____

_photosynthesis_

_____

_____

**Academic Vocabulary**

**Define** convert *using a dictionary. Then use the word in a sentence to illustrate its scientific meaning.*

_convert_

_____

_____

Copyright © Glencoe/McGraw-Hill, a division of The McGraw-Hill Companies, Inc.

## Section 1 The World of Cells (continued)

**Main Idea**  |  **Details**

**Importance of Cells**

*I found this information on page _____.*

**Summarize** *the three main ideas of cell theory.*

| | Cell Theory |
|---|---|
| 1. | All living things are made up of one or more cells. |
| 2. | |
| 3. | |

**What are cells made of?**

*I found this information on page _____.*

**Organize** *information you have learned about parts of a cell.*

Parts of a cell

I. The outside of the cell

A. _____ (plants only)

supports and _____

B. cell membrane

1. _____

2. _____

II. The inside of the cell

A. _____

1. gelatin-like substance

2. _____

B. _____

1. _____

a. stores _____ in chromosomes

b. _____

2. Vacuoles store _____, _____,

_____, and _____

3. _____ converts food energy into

_____

Copyright © Glencoe/McGraw-Hill, a division of The McGraw-Hill Companies, Inc.

## Section 1 The World of Cells (continued)

**Main Idea**      **Details**

### What are cells made of?

*I found this information on page _____ .*

**Model** *an animal cell. Use your book to help you sketch an animal cell and label its parts.*

    cell membrane      cytoplasm      nucleus

    chromosomes      mitochondrion      vacuole

### Energy and the Cell

*I found this information on page _____ .*

**Compare** *cellular respiration and* **photosynthesis.** *Label each input and output flow chart with these same five labels.*

    carbon dioxide      food      energy      oxygen      water

#### Cellular Respiration

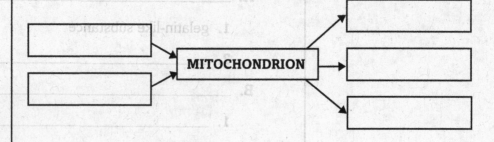

*I found this information on page _____ .*

#### Photosynthesis

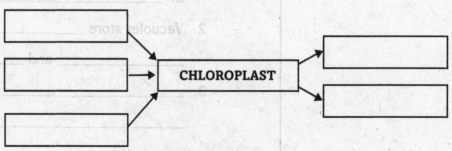

Copyright © Glencoe/McGraw-Hill, a division of The McGraw-Hill Companies, Inc.

# Cells—The Units of Life

## Section 2  The Different Jobs of Cells

**Skim** *the section. Read the headings and the figure captions. Predict three topics that might be discussed in this section.*

1. _____

2. _____

3. _____

### Review Vocabulary

**Define** organism *using a dictionary.*

organism _____

_____

### New Vocabulary

*Read the definitions below. Write the key term on the blank in the left column.*

_____  groups of similar cells that do the same type of work

_____  different types of tissues working together

_____  a group of organs that works together to do a certain job

### Academic Vocabulary

*Use a dictionary to define* function. *Then use the term in a scientific sentence.*

function _____

_____

_____

Copyright © Glencoe/McGraw-Hill, a division of The McGraw-Hill Companies, Inc.

## Section 2  The Different Jobs of Cells (continued)

| Main Idea | Details |
|---|---|

**Special Cells for Special Jobs**

*I found this information on page _____.*

**Summarize** *information from your book about human cells.*

| Type of Cell | Description |
|---|---|
| Bone | |
| | long and have many branches to send and receive messages quickly |
| | usually long and have many fibers that can contract and relax |
| Skin | |
| Fat | |

*I found this information on page _____.*

**Identify** *3 functions of plant cells.*

1. _____    3. _____

2. _____

*I found this information on page _____.*

**Compare and Contrast** *human skin cells and the cells on the outside of a plant stem. Put the statements into the Venn diagram.*

• cells are flat and close together
• part of the outer layer of the organism
• cells are short and thick

• provide protection against sun and disease
• cells provide structure

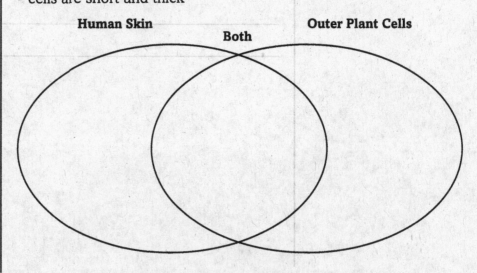

Human Skin          Both          Outer Plant Cells

Copyright © Glencoe/McGraw-Hill, a division of The McGraw-Hill Companies, Inc.

## Section 2  The Different Jobs of Cells (continued)

### Main Idea

**Cell Organization**

*I found this information on page _____ .*

### Details

**Organize** *information about cell organization by completing the outline.*

Cell organization of many-celled organisms

I. Tissues
  A. Definition: _____

  B. Example: _____

II. Organs
  A. Definition: _____

  B. Example: _____

  Specific examples of tissue system

    1. _____

    2. _____

    3. _____

III. Organ systems
  A. Definition: _____

       _____

  B. Example: _____

  Specific examples of organs in system

    1. _____

    2. _____

    3. _____

**CONNECT IT**  Create an analogy between the jobs of nerve cells and fat cells to real-life careers. For example, skin cells help protect the body, and police officers help protect people.

_____

_____

Copyright © Glencoe/McGraw-Hill, a division of The McGraw-Hill Companies, Inc.

# Cells—The Units of Life  Chapter Wrap-Up

*Now that you have read the chapter, think about what you have learned and complete the table below. Compare your previous answers with these.*

**1.** Write an **A** if you agree with the statement.

**2.** Write a **D** if you disagree with the statement.

| Cells—The Units of Life | After You Read |
|---|---|
| • Bacteria are the smallest organisms on Earth. | |
| • All living things are made up of one or more cells. | |
| • A cell's shape and size can be related to its function. | |
| • Cells are organized into systems to perform functions that keep an organism alive. | |

# Review

*Use this checklist to help you study.*

☐ Review the information you included in your Foldable.

☐ Study your *Science Notebook* on this chapter.

☐ Study the definitions of vocabulary words.

☐ Review daily homework assignments.

☐ Re-read the chapter and review the charts, graphs, and illustrations.

☐ Review the Self Check at the end of each section.

☐ Look over the Chapter Review at the end of the chapter.

## SUMMARIZE IT

After reading this chapter, identify three things that you have learned about cells.

_____

_____

_____

_____

Copyright © Glencoe/McGraw-Hill, a division of The McGraw-Hill Companies, Inc.

# Invertebrate Animals

## Before You Read

*Before you read the chapter, respond to these statements.*

1. Write an **A** if you agree with the statement.
2. Write a **D** if you disagree with the statement.

| Before You Read | Invertebrate Animals |
|---|---|
| | • Most animals have a backbone. |
| | • Animals are made up of many cells and have many different types of cells. |
| | • Animals can make their own food. |
| | • All animals can digest their food. |
| | • All animals can move from place to place. |

**FOLDABLES**
**Study Organizer**

*Construct the Foldable as directed at the beginning of this chapter.*

**Science Journal**

*Describe similarities and differences between you and an aquatic invertebrate animal such as a nudibranch, which is a type of sea slug.*

_____

_____

_____

_____

_____

Copyright © Glencoe/McGraw-Hill, a division of The McGraw-Hill Companies, Inc.

# Invertebrate Animals
## Section 1  What is an animal?

**Preview** *Section 1 by reading the headings. Write three questions you have about the content of the section.*

1. _____

2. _____

3. _____

### Review Vocabulary

**Define** organelle *using your book or a dictionary.*

organelle

_____

_____

### New Vocabulary

*Define the following key terms. Below each definition, copy one sentence from Section 1 of your book that uses the word. Do not copy the sentence that gives the definition.*

symmetry

_____

_____

_____

invertebrate

_____

_____

_____

### Academic Vocabulary

*Use a dictionary to define* indicate.

indicate

_____

_____

_____

Copyright © Glencoe/McGraw-Hill, a division of The McGraw-Hill Companies, Inc.

Section 1  What is an animal? (continued)

## Main Idea

## Details

### Animal Characteristics

*I found this information on page _____.*

**Complete** *the following chart by writing a statement about each characteristic of animals.*

| Animals | |
|---|---|
| **Characteristic** | **Statement** |
| Cells | |
| Nucleus and organelles | |
| Obtaining energy | |
| Digesting food | |
| Movement | |

### Symmetry

*I found this information on page _____.*

**Compare** *forms of animal symmetry by drawing an example for each of the three types of symmetry below.*

| **Asymmetry** | **Bilateral Symmetry** | **Radial Symmetry** |
|---|---|---|
| | | |

Copyright © Glencoe/McGraw-Hill, a division of The McGraw-Hill Companies, Inc.

## Section 1  What is an animal? (continued)

**Main Idea**

**Details**

### Animal Classification

*I found this information on page _____.*

**Classify** *the types of* invertebrates *in the chart below.*

| Animal Kingdom | | |
|---|---|---|
| Invertebrates | | |
| | | |
| | | |
| | | |

---

**CONNECT IT**  Design an imaginary animal species. Keep in mind the five common characteristics of animals. Give your animal species a name. Draw it and label its parts.

**My animal species:** _____

Copyright © Glencoe/McGraw-Hill, a division of The McGraw-Hill Companies, Inc.

# Invertebrate Animals
## Section 2  Sponges, Cnidarians, Flatworms, and Roundworms

**Scan** *the figures in Section 2 of your book. Write three questions that come to your mind.*

1. _____

_____

2. _____

_____

3. _____

_____

**Review Vocabulary**  **Define** *species to show its scientific meaning.*

species  _____

_____

**New Vocabulary**  *Use your book to define the following key terms.*

cnidarian  _____

_____

polyp  _____

_____

medusa  _____

_____

**Academic Vocabulary**  *Use your book or a dictionary to find two meanings for the term segment. Write both definitions below.*

segment  _____

_____

Copyright © Glencoe/McGraw-Hill, a division of The McGraw-Hill Companies, Inc.

## Section 2 Sponges, Cnidarians, Flatworms, and Roundworms (continued)

| Main Idea | Details |
|---|---|

**Sponges**

*I found this information on page _____.*

**Organize** *the information about sponges by filling in the key information.*

**A.** Filter feeders

_____

_____

**B.** Body support and defense

_____

_____

**C.** Sponge reproduction

_____

_____

**Cnidarians**

*I found this information on page _____.*

**Compare** *the two body forms of cnidarians by describing them in words and by drawing them in the chart below.*

| Cnidarian Body Forms | | |
|---|---|---|
| Form | Description | Drawing |
| Polyp | | |
| Medusa | | |

Copyright © Glencoe/McGraw-Hill, a division of The McGraw-Hill Companies, Inc.

Section 2  Sponges, Cnidarians, Flatworms, and Roundworms (continued)

## Main Idea

## Details

### Cnidarian Reproduction

*I found this information on page _____.*

**Sequence** *the main stages of reproduction in* medusa *forms of cnidarian, starting and ending with larva. Refer to the life cycle diagram in your book if you need help.*

1. _____

2. _____

3. _____

4. _____

5. _____

### Flatworms And Roundworms

*I found this information on page _____.*

**Compare and contrast** *characteristics of flatworms and roundworms by completing the chart below.*

|  | **Flatworms** | **Roundworms** |
|---|---|---|
| Body shape |  |  |
| Body openings |  |  |
| Body construction |  |  |
| Digestive system |  |  |

### CONNECT IT

Evaluate how the ability to move from place to place would give an invertebrate an advantage in getting food and reproducing.

_____

_____

Copyright © Glencoe/McGraw-Hill, a division of The McGraw-Hill Companies, Inc.

# Invertebrate Animals
## Section 3  Mollusks and Segmented Worms

**Scan** *Section 3 of your textbook. Then write two facts that you learned about mollusks and segmented worms.*

1. _____

2. _____

**Review Vocabulary**  **Define** *organ using your book as it applies to living organisms.*

organ  _____

_____

**New Vocabulary**  *Define the following key terms.*

mollusk  _____

mantle  _____

_____

radula  _____

_____

open circulatory system  _____

_____

closed circulatory system  _____

_____

**Academic Vocabulary**  *Use a dictionary to define the word* rigid.

rigid  _____

_____

Copyright © Glencoe/McGraw-Hill, a division of The McGraw-Hill Companies, Inc.

## Section 3 Mollusks and Segmented Worms (continued)

**⟨ Main Idea ⟩**           **⟨ Details ⟩**

### Mollusks

*I found this information on page _____.*

**Organize** *the information in your book by writing the six important characteristics of* mollusks.

1. _____
2. _____
3. _____
4. _____
5. _____
6. _____

### Types of Mollusks

*I found this information on page _____.*

**Classify** *the three types of mollusks by completing the chart below.*

| Mollusks | | | |
|---|---|---|---|
| Types | | | |
| Where do they live? | | | |
| How many shells? | | | |
| Examples | | | |

### Cephalapods

*I found this information on page _____.*

**Describe** *the movement of a squid in water. Refer to the drawing of a balloon in your book if you need help.*

_____

_____

_____

_____

_____

Copyright © Glencoe/McGraw-Hill, a division of The McGraw-Hill Companies, Inc.

## Section 3 Mollusks and Segmented Worms (continued)

| ⟨Main Idea⟩ | ⟨Details⟩ |
|---|---|

**Segmented Worms**

*I found this information on page _____.*

**Summarize** *the four characteristics of segmented worms below.*

1. _____
2. _____
3. _____
4. _____

**Types of Segmented Worms**

*I found this information on page _____.*

**Classify** *types of segmented worms by completing the chart.*

| Types of Segmented Worms | | | |
|---|---|---|---|
| Type | Where Found | Source of Energy | An Interesting Characteristic |
| Earthworm | | | |
| Leech | | | |
| Marine worm | | | |

**CONNECT IT** Write an account of an hour in the life of an earthworm. Include information about how the worm moves and eats.

_____

_____

_____

Copyright © Glencoe/McGraw-Hill, a division of The McGraw-Hill Companies, Inc.

# Invertebrate Animals
## Section 4  Arthropods and Echinoderms

**Scan** *the illustrations in this section. Write four things you learned about arthropods and echinoderms from the illustrations.*

1. _____

2. _____

3. _____

4. _____

**Review Vocabulary**  **Define** regeneration *using your book or a dictionary.*

regeneration _____

_____

**New Vocabulary**  *Define the following vocabulary terms.*

arthropod _____

_____

appendage _____

_____

exoskeleton _____

_____

metamorphosis _____

_____

**Academic Vocabulary**  *Use your book or a dictionary to define* inject. *Use the word in a sentence about how spiders capture prey.*

inject _____

_____

Copyright © Glencoe/McGraw-Hill, a division of The McGraw-Hill Companies, Inc.

## Section 4 Arthropods and Echinoderms (continued)

| ⬭ Main Idea ⬭ | ⬭ Details ⬭ |
|---|---|

**Arthropods**

*I found this information on page _____.*

**Organize** *information from your book by filling in the web diagram with the five characteristics shared by all* **arthropods.**

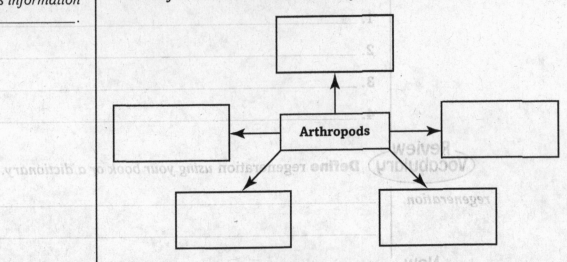

Arthropods

*I found this information on page _____.*

**Analyze** *the information in your book to complete the following chart about the four types of arthropods.*

| Types of Arthropods | |
|---|---|
| **Type** | **Characteristics** |
| Insects | |
| Arachnids | |
| Centipedes and millipedes | |
| Crustaceans | |

Copyright © Glencoe/McGraw-Hill, a division of The McGraw-Hill Companies, Inc.

Section 4  Arthropods and Echinoderms (continued)

### Main Idea

*I found this information on page _____.*

### Details

**Sequence** *the stages of complete and incomplete* **metamorphosis** *by labeling the charts.*

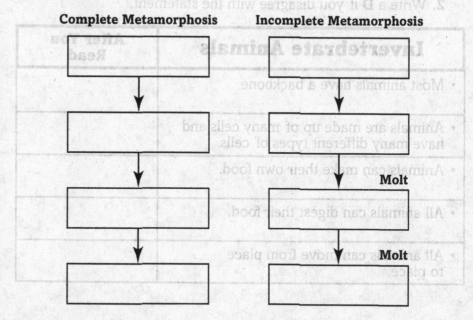

Complete Metamorphosis          Incomplete Metamorphosis

Molt

Molt

### Echinoderms

*I found this information on page _____.*

**Summarize** *characteristics common to echinoderms by making a list of characteristics below.*

1. _____

2. _____

3. _____

4. _____

**CONNECT IT** Compare the circulatory systems of an insect and an earthworm.

_____

_____

_____

_____

_____

Copyright © Glencoe/McGraw-Hill, a division of The McGraw-Hill Companies, Inc.

# Invertebrate Animals Chapter Wrap-Up

*Now that you have read the chapter, think about what you have learned and complete the table below. Compare your previous answers with these.*

1. Write an **A** if you agree with the statement.
2. Write a **D** if you disagree with the statement.

| Invertebrate Animals | After You Read |
|---|---|
| • Most animals have a backbone. | |
| • Animals are made up of many cells and have many different types of cells. | |
| • Animals can make their own food. | |
| • All animals can digest their food. | |
| • All animals can move from place to place. | |

## Review
*Use this checklist to help you study.*

- ☐ Review the information you included in your Foldable.
- ☐ Study your *Science Notebook* on this chapter.
- ☐ Study the definitions of vocabulary words.
- ☐ Review daily homework assignments.
- ☐ Re-read the chapter and review the charts, graphs, and illustrations.
- ☐ Review the Self Check at the end of each section.
- ☐ Look over the Chapter Review at the end of the chapter.

**SUMMARIZE IT** After reading this chapter, identify three things that you have learned about invertebrate animals.

_____

_____

_____

_____

Copyright © Glencoe/McGraw-Hill, a division of The McGraw-Hill Companies, Inc.

# Vertebrate Animals

## Before You Read

*Before you read the chapter, think about what you know about the topic. List three things that you already know about animals with backbones in the first column. Then list three things that you would like to learn about them in the second column.*

| K<br>What I know | W<br>What I want to find out |
|---|---|
|  |  |
|  |  |
|  |  |
|  |  |

*Construct the Foldable as directed at the beginning of this chapter.*

**Science Journal**

*An eagle, a salmon, a snake, and a grizzly bear all have a backbone. List other traits these animals and humans share.*

_____

_____

_____

_____

_____

_____

Copyright © Glencoe/McGraw-Hill, a division of The McGraw-Hill Companies, Inc.

# Vertebrate Animals

## Section 1 Chordate Animals

**Skim** *the headings in Section 3. Then make three predictions about what you will learn.*

1. _____

2. _____

3. _____

**Review Vocabulary**  **Define** *Write the correct word next to each definition.*

_____  an animal without a backbone

**New Vocabulary**

_____  an animal with a notochord, a nerve cord, and pharyngeal pouches sometime during development

_____  a vertebrate whose body temperature changes as the surrounding temperature changes

_____  an animal whose body temperature does not change with changes in the surrounding temperature

_____  a tough flexible tissue that is similar to bone but not as hard or brittle

**Academic Vocabulary**  *Use a dictionary to define* maintain.

*maintain*  _____

_____

Copyright © Glencoe/McGraw-Hill, a division of The McGraw-Hill Companies, Inc.

Name _____  Date _____

## Section 1 Chordate Animals (continued)

⟨ **Main Idea** ⟩                    ⟨ **Details** ⟩

**What is a chordate?**

*I found this information on page _____.*

**Identify** *and describe three characteristics of all chordates that appear at some time during their development.*

| Chordates | | |
|---|---|---|
| **Characteristic** | | **Description** |
| **1.** | | |
| **2.** | | |
| **3.** | | |

*I found this information on page _____.*

**Model** *a simple chordate by copying the sketch of the lancelet. Sketch a human next to it.*

*Name each human structure with the same function as the following lancelet structures.*

• notochord: _____

• nerve cord: _____

• gill slit: _____

*I found this information on page _____.*

**Compare** *the characteristics that all chordates share to the characteristics that only vertebrates share.*

| **All Chordates** | **Only Vertebrates** |
|---|---|
| | |

Copyright © Glencoe/McGraw-Hill, a division of The McGraw-Hill Companies, Inc.

Section 1 Chordate Animals (continued)

**Main Idea**                **Details**

**Fish and Types of Fish**

I found this information
on page _____.

**Contrast** *the characteristics of bony fish, jawless fish, and cartilaginous fish by completing the diagram. Write 3–4 characteristics for each type.*

| Characteristics of All Fish | | |
|---|---|---|
| Bony Fish | Jawless Fish | Cartilaginous Fish |
|  |  |  |

I found this information
on page _____.

**Analyze** *the adaptations of a typical bony fish. Use the figure in your book to help you sketch and label the fish.*

**CONNECT IT** Compare ectotherms and endotherms. Hypothesize about the advantages and disadvantages of each.

_____

_____

_____

_____

Copyright © Glencoe/McGraw-Hill, a division of The McGraw-Hill Companies, Inc.

# Vertebrate Animals
## Section 2 Amphibians and Reptiles

**Scan** *Section 1 of your book. Then write three questions that you have about amphibians and reptiles. Try to answer your questions as you read.*

1. _____

2. _____

3. _____

**Review Vocabulary** **Define** metamorphosis *to show its scientific meaning.*

metamorphosis _____

_____

**New Vocabulary** *Use a dictionary or your book to define each key term.*

hibernation _____

_____

estivation _____

_____

amniotic egg _____

_____

**Academic Vocabulary** *Use a dictionary to define* internal.

internal _____

_____

Copyright © Glencoe/McGraw-Hill, a division of The McGraw-Hill Companies, Inc.

## Section 2 Amphibians and Reptiles (continued)

| Main Idea | Details |
|---|---|

**Amphibians**

*I found this information on page* _____.

**Complete** *the prompts about amphibians.*

Definition: _____

_____

Origin of the word *amphibian:* _____

_____

Examples: _____

_____

*I found this information on page* _____.

**Contrast** *hibernation and estivation in amphibians by completing the Venn diagram with at least five facts.*

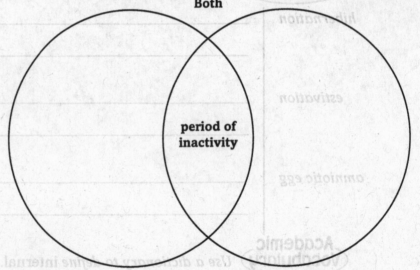

Hibernation          Estivation

Both

period of inactivity

*I found this information on page* _____.

**Organize** *amphibian characteristics by listing them below.*

1. _____

2. _____

3. _____

4. _____

5. _____

Copyright © Glencoe/McGraw-Hill, a division of The McGraw-Hill Companies, Inc.

## Section 2 Amiphibians and Reptiles (continued)

| Main Idea | Details |
|---|---|

### Reptiles

*I found this information on page _____.*

**Organize** *information about reptiles by completing the diagram.*

| Turtles | Crocodiles and Alligators |
|---|---|
|  |  |
| **Lizards** | **Snakes** |
|  |  |

*I found this information on page _____.*

**Summarize** *adaptations that are typical of reptiles by completing the chart.*

| Reptile Adaptations | |
|---|---|
| Characteristic | Purpose |
| Skin |  |
| Neck |  |
| Lungs |  |
| Internal fertilization |  |
| Amniotic eggs |  |

### CONNECT IT

Compare and contrast amphibians and reptiles.

_____

_____

_____

_____

Copyright © Glencoe/McGraw-Hill, a division of The McGraw-Hill Companies, Inc.

# Vertebrate Animals
## Section 3 Birds

**Skim** the headings in Section 3. Then make 3 predictions about what you will learn.

1. _____

2. _____

3. _____

### Review Vocabulary

**Define** appendage *to show its scientific meaning. Think of two examples of appendages.*

appendage _____

### New Vocabulary

*Use your book to define and sketch each type of feather.*

contour feather _____

down feather _____

### Academic Vocabulary

*Use a dictionary to define* constant.

constant _____

Copyright © Glencoe/McGraw-Hill, a division of The McGraw-Hill Companies, Inc.

Section 3  Birds (continued)

## Main Idea        ## Details

| Characteristics of Birds | List *six characteristics of birds.* |
|---|---|
| *I found this information on page* _____. | 1. _____ |
| | 2. _____ |
| | 3. _____ |
| | 4. _____ |
| | 5. _____ |
| | 6. _____ |

| Adaptations for Flight | Analyze *how birds are adapted for flight. Make a concept web that includes five adaptations.* |
|---|---|
| *I found this information on page* _____. | |

Copyright © Glencoe/McGraw-Hill, a division of The McGraw-Hill Companies, Inc.

Section 3  Birds (continued)

<Main Idea>　　　　　　　　　<Details>

**Functions of Feathers**

*I found this information on page _____.*

**Compare and contrast** *contour feathers and down feathers. List characteristics of each type of feather.*

| Down Feathers | Contour Feathers |
|---|---|
| 1. | 1. |
| 2. | 2. |
| 3. | 3. |
| 4. | 4. |

*I found this information on page _____.*

**Analyze** *at least three reasons why birds preen.*

1. _____

_____

2. _____

_____

3. _____

_____

**COMPARE IT** Analyze which would be warmer: a winter coat stuffed with down feathers, or one made of woven cloth. Provide reasons to support your answer.

_____

_____

_____

_____

_____

Copyright © Glencoe/McGraw-Hill, a division of The McGraw-Hill Companies, Inc.

# Vertebrate Animals
## Section 4  Mammals

**Skim** *Section 4, then write four topics about mammals that you would like to know about.*

1. _____

2. _____

3. _____

4. _____

*Write the correct key word next to each definition.*

_____  the arrangement of the individual parts of an object that can be divided into similar parts

_____  plant-eating mammal with incisors specialized to cut vegetation and large, flat molars to grind it

_____  meat-eating animal with sharp canine teeth specialized to rip and tear flesh

_____  plant- and meat-eating animal with incisors that cut vegetables, sharp premolars that chew meat, and molars that grind food

_____  mammal whose offspring develops inside the female's uterus; has a placenta that supplies the embryo with food and oxygen and removes waste

_____  mammal that gives birth to incompletely developed young that finish developing in their mother's pouch

_____  mammal that lays eggs with tough, leathery shells instead of giving birth to live young

**Academic Vocabulary**  *Use a dictionary to define* complex.

*complex*  _____

_____

Copyright © Glencoe/McGraw-Hill, a division of The McGraw-Hill Companies, Inc.

Section 4 Mammals (continued)

## Main Idea

## Details

**Mammal Characteristics**

*I found this information on page _____.*

**Organize** *7 characteristics common to mammals.*

1. _____

2. _____

3. _____

4. _____

5. _____

6. _____

7. _____

*I found this information on page _____.*

**Model** *and describe the different kinds of teeth carnivores, omnivores, and herbivores have. Use the figure in your book to help you.*

Copyright © Glencoe/McGraw-Hill, a division of The McGraw-Hill Companies, Inc.

## Section 4  Mammals (continued)

Copyright © Glencoe/McGraw-Hill, a division of The McGraw-Hill Companies, Inc.

**Main Idea**

**Details**

### Mammal Types

*I found this information on page _____.*

**Classify** *mammals by completing the following chart.*

| Types of Mammals | | |
|---|---|---|
| Type of Mammal | Characteristics | Examples |
| Monotreme | | |
| Marsupial | | |
| Placental | | |

**CONNECT IT** Choose a wild mammal that is native to your area. Classify it using the information you have learned. Provide two unique characteristics. Tell how it is adapted to its environment.

_____

_____

_____

_____

# Vertebrate Animals  Chapter Wrap-Up

*Review the ideas you listed in the K-W-L chart at the beginning of the chapter. Cross out any incorrect information in the first column. Then complete the chart by filling in the third column.*

| K<br>What I know | W<br>What I want to find out | L<br>What I learned |
|---|---|---|
|  |  |  |

## Review

*Use this checklist to help you study.*

☐ Review the information you included in your Foldable.

☐ Study your *Science Notebook* on this chapter.

☐ Study the definitions of vocabulary words.

☐ Review daily homework assignments.

☐ Re-read the chapter and review the charts, graphs, and illustrations.

☐ Review the Self Check at the end of each section.

☐ Look over the Chapter Review at the end of the chapter.

---

## SUMMARIZE IT

After reading this chapter, identify three facts that you have learned about vertebrate animals.

_____

_____

_____

_____

Copyright © Glencoe/McGraw-Hill, a division of The McGraw-Hill Companies, Inc.

# The Human Body

## Before You Read

*Before you read the chapter, respond to these statements.*

1. Write an **A** if you agree with the statement.
2. Write a **D** if you disagree with the statement.

| Before You Read | The Human Body |
|---|---|
| | • Skin is the largest body organ. |
| | • Water is a nutrient. |
| | • Food is absorbed in the small intestine. |
| | • A human baby develops in about 28 days. |

**FOLDABLES™ Study Organizer**

*Construct the Foldable as directed at the beginning of this chapter.*

**Science Journal**

*Write three things your body needs to keep you healthy.*

_____

_____

_____

_____

_____

_____

Copyright © Glencoe/McGraw-Hill, a division of The McGraw-Hill Companies, Inc.

# The Human Body
Section 1  Body Systems

**Scan** *the headings in Section 1. Write three questions that come to mind from your reading.*

1. _____

2. _____

3. _____

**Review Vocabulary**  **Define** *the following terms using your book or a dictionary.*

organ _____

**New Vocabulary**

skeletal system _____

melanin _____

muscle _____

nutrient _____

respiratory system _____

alveoli _____

capillary _____

reflex _____

**Academic Vocabulary**  *Explain how the word* framework *relates to the human body.*

framework _____

Copyright © Glencoe/McGraw-Hill, a division of The McGraw-Hill Companies, Inc.

## Section 1  Body Systems (continued)

### Main Idea

### Details

#### Structure and Movement

*I found this information on page _____.*

**Organize** *5 important facts about the skeletal system.*

1. _____

2. _____

3. _____

4. _____

5. _____

*I found this information on page _____.*

**Create** *a concept map about the functions of skin. Include at least 6 pieces of information about the 4 important functions of this organ.*

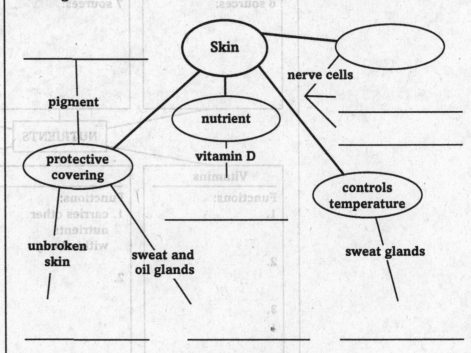

*I found this information on page _____.*

**Compare** *and contrast your body's two types of muscles.*

| | Voluntary Muscles | Involuntary Muscles |
|---|---|---|
| How they work | | |
| Examples | | |

Copyright © Glencoe/McGraw-Hill, a division of The McGraw-Hill Companies, Inc.

## Section 1  Body Systems (continued)

**Main Idea**                                   **Details**

### Digestion and Excretion

*I found this information on page _____.*

**Organize** *information about the 6 groups of nutrients in the concept web. Name each nutrient and list its functions and food sources (where asked).*

| Proteins | Function: | Functions: |
|---|---|---|
| **Functions:** | 1. **main supplier of energy for the body** | 1. |
| 1. | | |
| 2. | | 2. |
| | | 3. **cushion internal organs** |
| **6 sources:** | **7 sources:** | **4 sources:** |

**NUTRIENTS**

| Vitamins | Functions: | Functions: |
|---|---|---|
| **Functions:** | 1. **carries other nutrients within body** | 1. |
| 1. | | |
| 2. | 2. | 2. **make and maintain bone** |
| 3. | | |

*I found this information on page _____.*

**Identify** *the two groups of vitamins and describe how they differ.*

1. _____

_____

2. _____

Copyright © Glencoe/McGraw-Hill, a division of The McGraw-Hill Companies, Inc.

Section 1 Body Systems (continued)

<Main Idea>     <Details>

**Respiration and Circulation**

*I found this information on page _____.*

**Create** *a sequence diagram to show how oxygen travels through the respiratory system.*

*I found this information on page _____.*

**Identify** *the three types of blood vessels in the cycle diagram.*

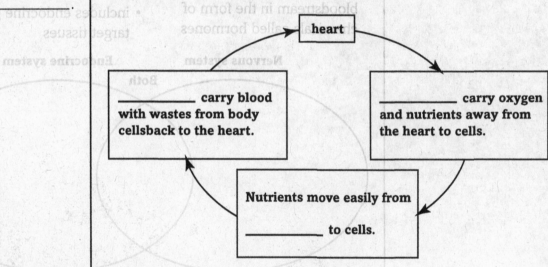

heart

_____ carry blood with wastes from body cellsback to the heart.

_____ carry oxygen and nutrients away from the heart to cells.

Nutrients move easily from _____ to cells.

*I found this information on page _____.*

**Organize** *information about blood by completing the chart.*

| Blood | |
|---|---|
| **Blood Part** | **Function** |
| Red cells | |
| White cells | |
| Platelets and blood chemicals | |

Copyright © Glencoe/McGraw-Hill, a division of The McGraw-Hill Companies, Inc.

## Section 1  Body Systems (continued)

| ⟨ **Main Idea** ⟩ | ⟨ **Details** ⟩ |
|---|---|

**Respiration and Circulation**

*I found this information on page _____ .*

**Analyze** *why blood must be sorted by type when it is donated.*

_____

_____

_____

_____

_____

_____

**Control and Coordination**

*I found this information on page _____ .*

**Compare** *the parts and function of the body's control systems by completing the Venn diagram using the following:*

- sends messages along nerve cells called neurons
- sends messages through the bloodstream in the form of chemicals called hormones

- send messages within the body
- includes brain, spinal cord, nerves, nerve receptors
- includes endocrine glands and target tissues

Nervous system          Endocrine system
**Both**

⬭⬭

┌─────────────────────────────────────────┐
│ **SYNTHESIZE IT** |
│ **Analogy** Imagine your body is a factory and that your
│ cells and body systems are workers. Describe the work at least three different types
│ of workers do in your body factory.
│
│ _____
│
│ _____
│
│ _____
└─────────────────────────────────────────┘

Copyright © Glencoe/McGraw-Hill, a division of The McGraw-Hill Companies, Inc.

# The Human Body

## Section 2 Human Reproduction

**Preview** *Section 2 by reading each* **What You'll Learn** *statement. Then, rewrite each statement so that it is a question. Try to answer these questions as you study the Section 2.*

1. _____

2. _____

3. _____

4. _____

**Review Vocabulary**) **Define** *the following terms by writing the term next to its definition.*

_____  chemical produced by the endocrine system and released directly into the bloodstream

**New Vocabulary**)

_____  male reproductive cells

_____  mixture of fluid and sperm

_____  release of an egg from an ovary

_____  monthly cycle in a sexually mature female

_____  period of development from fertilized egg to birth

_____  zygote after it has attached to the wall of the uterus

_____  developing embryo after the first two months

**Academic Vocabulary**) *Use a dictionary to define* phase.

*phase* _____

_____

Copyright © Glencoe/McGraw-Hill, a division of The McGraw-Hill Companies, Inc.

## Section 2  Human Reproduction (continued)

| Main Idea | Details |
|---|---|

**Male Reproductive System** and **Female Reproductive System**

*I found this information on page* _____.

**Identify** *the organs of the male and female reproductive systems.*

**Male Reproductive System**

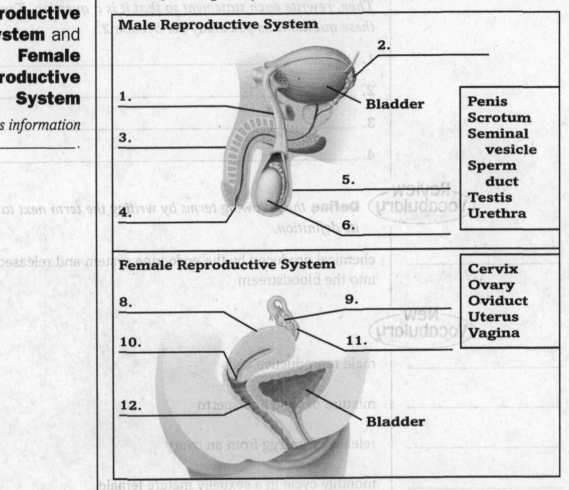

1. _____

2. _____

3. _____

Bladder

5. _____

4. _____

6. _____

| Penis |
| Scrotum |
| Seminal vesicle |
| Sperm duct |
| Testis |
| Urethra |

**Female Reproductive System**

8. _____

9. _____

10. _____

11. _____

12. _____

Bladder

| Cervix |
| Ovary |
| Oviduct |
| Uterus |
| Vagina |

*I found this information on page* _____.

**Describe** *the phases of the average menstrual cycle.*

Phase 1, _____ usually lasts 4–6 days.

_____

Phase 2, _____

_____. This phase ends with _____

Phase 3 can proceed in two different ways. If the egg is fertilized,

_____

If the egg is not fertilized, _____

_____

Copyright © Glencoe/McGraw-Hill, a division of The McGraw-Hill Companies, Inc.

Section 2 Human Reproduction (continued)

## Main Idea

### Life Stages

*I found this information on page _____.*

## Details

**Summarize** *the development of a baby before birth.*

### Development Through Birth

| Stage | Description |
|-------|-------------|
| Zygote | |
| Embryo | |
| Fetus | |
| Birth | |

*I found this information on page _____.*

**Identify** *the stages of development and write a sentence that describes major changes that take place at each stage.*

### Life Stages

| | |
|---------|---|
| Stage 1 | |
| Stage 2 | |
| Stage 3 | |
| Stage 4 | |

**CONNECT IT** Analyze physical changes you notice happening during various stages of life from infancy to late adulthood.

_____

_____

_____

Copyright © Glencoe/McGraw-Hill, a division of The McGraw-Hill Companies, Inc.

# The Human Body Chapter Wrap-Up

*Now that you have read the chapter, think about what you have learned and complete the table below. Compare your previous answers with these.*

1. Write an **A** if you agree with the statement.

2. Write a **D** if you disagree with the statement.

| The Human Body | After You Read |
|---|---|
| • Skin is the largest body organ. | |
| • Water is a nutrient. | |
| • Food is absorbed in the small intestine. | |
| • A human baby develops in about 28 days. | |

## Review

*Use this checklist to help you study.*

- ☐ Review the information you included in your Foldable.
- ☐ Study your *Science Notebook* on this chapter.
- ☐ Study the definitions of vocabulary words.
- ☐ Review daily homework assignments.
- ☐ Re-read the chapter and review the charts, graphs, and illustrations.
- ☐ Review the Self Check at the end of each section.
- ☐ Look over the Chapter Review at the end of the chapter.

### SUMMARIZE IT

After reading this chapter, identify three things that you have learned about the human body.

_____

_____

_____

_____

Copyright © Glencoe/McGraw-Hill, a division of The McGraw-Hill Companies, Inc.

# The Role of Genes in Inheritance

## Before You Read

*Before you read the chapter, respond to these statements.*

   **1.** Write an **A** if you agree with the statement.

   **2.** Write a **D** if you disagree with the statement.

| Before You Read | The Role of Genes in Inheritance |
|---|---|
| | • Offspring always show the dominant traits of their parents. |
| | • Some organisms can regrow parts of their bodies if these parts are lost. |
| | • Traits are passed from one generation to the next. |
| | • The environment cannot affect the way a person appears. |

**FOLDABLES™**
**Study Organizer**

*Construct the Foldable as directed at the beginning of this chapter.*

**Science Journal**

*Write three traits of horses that you could trace from parents to offspring.*

_____

_____

_____

_____

_____

_____

Copyright © Glencoe/McGraw-Hill, a division of The McGraw-Hill Companies, Inc.

# The Role of Genes in Inheritance

## Section 1  Continuing Life

**Skim** *the headings, illustrations, and charts in Section 1. Write three concepts that you predict this section will describe.*

1. _____

2. _____

3. _____

**Review Vocabulary**

**Define** chromosome *to show its scientific meaning.*

chromosome _____

_____

**New Vocabulary**

*Write sentences that contain both terms in each pair.*

asexual reproduction/ mitosis _____

_____

DNA/cloning _____

_____

sexual reproduction/ fertilization _____

_____

_____

meiosis/sex cells _____

_____

**Academic Vocabulary**

*Use your book or a dictionary to define the term* identical.

identical _____

_____

_____

Copyright © Glencoe/McGraw-Hill, a division of The McGraw-Hill Companies, Inc.

## Section 1 Continuing Life (continued)

| ~Main Idea~ | ~Details~ |
|---|---|

### Reproduction

*I found this information on page _____.*

**State** *two reasons that reproduction is important.*

1. _____

_____

2. _____

_____

*I found this information on page _____.*

**Complete** *the following paragraph.*

_____ is in all cells. It is shaped like a _____

_____. The sides support the steps, or rungs, of the

ladder. Each rung is made up of _____.

There are _____ bases, and they pair _____.

The order of the bases forms a _____ that provides the

cell with _____ about what materials to make, how to

make them, and when to make them.

### Cell Division

*I found this information on page _____.*

**Model** *the steps of mitosis and cell division, beginning with a cell that has four chromosomes. Then complete the caption below.*

In a plant or animal cell, cell division results in _____

and the _____ of aging or _____ cells.

Copyright © Glencoe/McGraw-Hill, a division of The McGraw-Hill Companies, Inc.

Section 1 Continuing Life (continued)

<Main Idea>                         <Details>

**Reproduction by One Organism**

I found this information on page _____ .

**Complete** *the information below about some important processes that rely on cell division.*

_____: Some organisms can replace body parts that have been lost.

Budding: _____

_____

_____: A copy of the original organism is made.

**Sex Cells and Reproduction, Production of Sex Cells, and Sex Cells in Plants**

I found this information on page _____ .

**Organize** *the information about* sex cells *by completing the outline.*

I. Types of human sex cells

A. _____: sperm

B. _____: _____

II. Production of sex cells

A. Sex cells are formed through _____ .

B. Sex cells have _____ the genetic information of

_____ .

III. Sex cells in flowering plants

A. After sperm and egg join, _____

_____ .

B. A _____ that contains _____ may then develop.

┌─────────────────────────────────────────────────────────────
│ **SYNTHESIZE IT**   Describe why it is important that sex cells are produced by
│ meiosis and not by mitosis.
│
│ _____
│
│ _____
│
│ _____
│
│ _____
│
│ _____
└─────────────────────────────────────────────────────────────

Copyright © Glencoe/McGraw-Hill, a division of The McGraw-Hill Companies, Inc.

# The Role of Genes in Inheritance
## Section 2  Genetics—The Study of Inheritance

**Scan** *Section 2. Read all of the section headings and bold terms. Write two facts that you discovered about genetics as you scanned the section.*

1. _____

2. _____

**Review Vocabulary**   **Define** *the term* genotype *to show its scientific meaning.*

genotype  _____

**New Vocabulary**   *Write the correct vocabulary word next to each definition.*

_____ passing of traits from parents to offspring

_____ study of how traits are passed from parents to offspring

_____ small section of DNA on a chromosome that has information about a trait

_____ different way that a certain trait appears that results from permanent changes in an organism's genes

_____ change in a gene or chromosome

**Academic Vocabulary**   **Define** feature *as it is used in the following sentence.*

Eye color, nose shape, and other features are traits that are inherited from one's parents.

feature  _____

_____

Copyright © Glencoe/McGraw-Hill, a division of The McGraw-Hill Companies, Inc.

Section 2 Genetics—The Study of Inheritance (continued)

**Main Idea** | **Details**

### Heredity

*I found this information on page* _____.

**Synthesize** *information about heredity by describing how traits are passed from parent to offspring.*

_____

_____

_____

_____

_____

### What determines traits?

*I found this information on page* _____.

**Analyze** hybrid *and* pure *traits by filling in the blanks.*

Each gene of a gene pair is called a(n) _____. If a gene pair contains different _____ for a trait, that trait is called a(n) _____. If a gene pair contains identical _____ for a trait, that trait is called _____.

*I found this information on page* _____.

**Identify** *whether the* dominant *or* recessive *form of the trait will be expressed in each case.*

| Alleles | Form of the Trait Expressed |
|---|---|
| two dominant alleles | . |
| one dominant allele, one recessive allele | |
| two recessive alleles | |

*I found this information on page* _____.

**Summarize** *how environment can affect the expression of traits.*

_____

_____

_____

_____

Copyright © Glencoe/McGraw-Hill, a division of The McGraw-Hill Companies, Inc.

## Section 2 Genetics—The Study of Inheritance (continued)

### Main Idea

### Details

**Passing Traits to Offspring**

*I found this information on page _____.*

**Analyze** *how a hybrid purple-flowered plant and a white-flowered plant can produce a purple-flowered plant. Fill in the correct allele(s) in each cell below.*

Purple-flowered
parent plant
sex cells

White-flowered
parent plant
sex cells

Offspring
cell

**Differences in Organisms**

*I found this information on page _____.*

**Complete** *the chart that shows causes of variation in a species.*

|  | Description | Example(s) |
|---|---|---|
| Multiple alleles | There are more than two alleles for a trait in a population. |  |
| Multiple genes |  |  |
| Mutations |  | four-leaf clover |

**SYNTHESIZE IT**

The allele that codes for the presence of dimples is a dominant allele. Explain why a girl might not have dimples even though both her parents have dimples.

_____

_____

_____

Copyright © Glencoe/McGraw-Hill, a division of The McGraw-Hill Companies, Inc.

# The Role of Genes in Inheritance
## Chapter Wrap-Up

*Now that you have read the chapter, think about what you have learned and complete the table below. Compare your previous answers with these.*

1. Write an **A** if you agree with the statement.
2. Write a **D** if you disagree with the statement.

| The Role of Genes in Inheritance | After You Read |
|---|---|
| • Offspring always show the dominant traits of their parents. | |
| • Some organisms can regrow parts of their bodies if these parts are lost. | |
| • Traits are passed from one generation to the next. | |
| • The environment cannot affect the way a person appears. | |

## Review

*Use this checklist to help you study.*

☐ Review the information you included in your Foldable.

☐ Study your *Science Notebook* on this chapter.

☐ Study the definitions of vocabulary words.

☐ Review daily homework assignments.

☐ Re-read the chapter and review the charts, graphs, and illustrations.

☐ Review the Self Check at the end of each section.

☐ Look over the Chapter Review at the end of the chapter.

### SUMMARIZE IT

**What are three important ideas in this chapter?**

_____

_____

_____

_____

Copyright © Glencoe/McGraw-Hill, a division of The McGraw-Hill Companies, Inc.

# Ecology

## Before You Read

*Before you read the chapter, respond to these statements.*

1. Write an **A** if you agree with the statement.
2. Write a **D** if you disagree with the statement.

| Before You Read | Ecology |
|---|---|
| | • The biosphere is made up of all of the ecosystems on Earth combined. |
| | • Different species of organisms live in the same habitat. |
| | • Energy for most organisms comes from the Sun. |
| | • A producer relies on prey for its energy. |

*Construct the Foldable as directed at the beginning of this chapter.*

**Science Journal**

*Describe how fallen leaves and insects contribute to the survival of frogs in a system.*

_____

_____

_____

_____

_____

_____

Copyright © Glencoe/McGraw-Hill, a division of The McGraw-Hill Companies, Inc.

# Ecology

## Section 1  What is an ecosystem?

**Skim** *Section 1. Predict three things that might be discussed in this section.*

1. _____

2. _____

3. _____

**Review Vocabulary**

**Define** organism *to show its scientific meaning.*

organism _____

_____

**New Vocabulary**

*Use your book to define the following key terms.*

ecosystem _____

ecology _____

_____

biotic factors _____

_____

abiotic factors _____

_____

**Academic Vocabulary**

*Use a dictionary to define* interact *to show its scientific meaning.*

interact _____

Copyright © Glencoe/McGraw-Hill, a division of The McGraw-Hill Companies, Inc.

## Section 1  What is an ecosystem? (continued)

<table>
<tr><td>

**Main Idea**

</td><td>

**Details**

</td></tr>
<tr><td>

### Ecosystems

*I found this information on page* _____.

</td><td>

**Identify** *some of the major ecosystems that make up the biosphere by completing the graphic organizer below.*

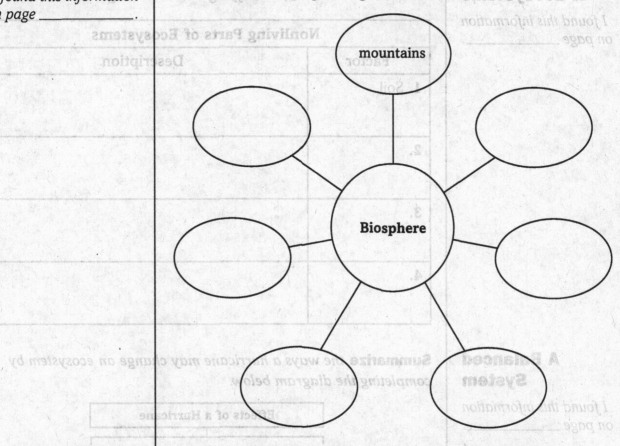

</td></tr>
<tr><td>

### Living Parts of Ecosystems

*I found this information on page* _____.

</td><td>

**Identify** *the four key needs of organisms and list them below.*

| Key Needs of Organisms |
| --- |
| 1. |
| 2. |
| 3. |
| 4. |

</td></tr>
</table>

Copyright © Glencoe/McGraw-Hill, a division of The McGraw-Hill Companies, Inc.

Name _____    Date _____

## Main Idea ~ Details

### Nonliving Parts of Ecosystems

I found this information on page _____.

**Organize** information about the four nonliving parts of ecosystems. Fill in the chart below, identifying and describing each.

| Nonliving Parts of Ecosystems | |
|---|---|
| Factor | Description |
| 1. Soil | |
| 2. | |
| 3. | |
| 4. | |

### A Balanced System

I found this information on page _____.

**Summarize** the ways a hurricane may change an ecosystem by completing the diagram below.

| Effects of a Hurricane | |
|---|---|
| Destructive | Beneficial |
| | |

**CONNECT IT** A fire sweeps through a forest ecosystem. Describe a destructive effect and a beneficial effect that may result.

_____

_____

_____

_____

Copyright © Glencoe/McGraw-Hill, a division of The McGraw-Hill Companies, Inc.

# Ecology
## Section 2  Relationships Among Living Things

**Skim** *Section 2 of your text. Write three questions that come to mind as you read the headings and examine the illustrations.*

1. _____

2. _____

3. _____

**Review Vocabulary**

**Define** *the following terms to show their scientific meanings.*

adaptation _____

**New Vocabulary**

population _____
_____

community _____

limiting factor _____
_____

niche _____

habitat _____
_____

**Academic Vocabulary**

*Use a dictionary to define* decline *to show its scientific meaning.*

decline _____
_____

Copyright © Glencoe/McGraw-Hill, a division of The McGraw-Hill Companies, Inc.

## Section 2 Relationships Among Living Things (continued)

### Main Idea

### Details

**Organizing Ecosystems**

*I found this information on page _____.*

**Complete** *the Venn diagram below to represent the relationship between a* population *and a* community.

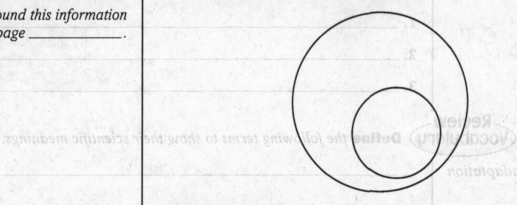

*I found this information on page _____.*

**Summarize** *the characteristics of populations that are studied by ecologists. Complete the sentence.*

The characteristics of a population include the size of the

population, _____,

and _____.

*I found this information on page _____.*

**Sequence** *the steps in the* mark and recapture method *of studying populations by completing the flow chart below.*

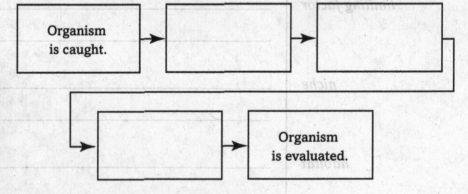

Populations can also be studied by:

1. _____

2. _____

3. _____

4. _____

Copyright © Glencoe/McGraw-Hill, a division of The McGraw-Hill Companies, Inc.

## Section 2  Relationships Among Living Things (continued)

<table>
<tr><td>Main Idea</td><td>Details</td></tr>
<tr><td>

### Limits to Populations

*I found this information on page _____.*

</td><td>

**Complete** *the graphic organizer below with some of the* resources *for which organisms compete.*

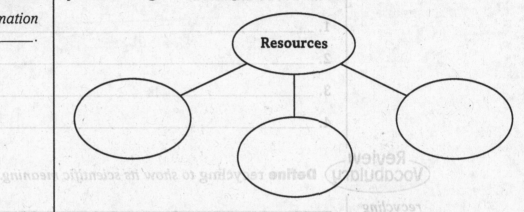

</td></tr>
<tr><td>

### Where and How Organisms Live

*I found this information on page _____.*

</td><td>

**Analyze** *the* niches *of snails, fish, and algae in an aquarium. Describe how each organism interacts with the other organisms and the environment.*

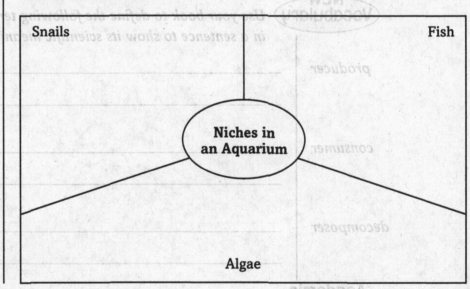

</td></tr>
</table>

## CONNECT IT

Describe how carpenter ants might both use resources and serve as a resource in the habitat of an apple tree.

_____

_____

_____

_____

Copyright © Glencoe/McGraw-Hill, a division of The McGraw-Hill Companies, Inc.

# Ecology

## Section 3 Energy Through the Ecosystem

**Scan** *the headings in Section 3 of your book. Identify four topics that will be discussed.*

1. _____

2. _____

3. _____

4. _____

### Review Vocabulary

**Define** recycling *to show its scientific meaning.*

recycling _____

_____

### New Vocabulary

*Use your book to define the following terms. Then use each term in a sentence to show its scientific meaning.*

producer _____

_____

consumer _____

_____

decomposer _____

_____

### Academic Vocabulary

*Use a dictionary to define* sequence *to show its scientific meaning.*

sequence _____

_____

_____

Copyright © Glencoe/McGraw-Hill, a division of The McGraw-Hill Companies, Inc.

## Section 3 Energy Through the Ecosystem (continued)

**Main Idea** · Details · **Details**

| **It's All About Food** and **Modeling The Flow of Energy** | **Organize** *the following to show relationships to one another in the flow of energy.* |

*I found this information on page _____.*

grasshopper          sunlight          insect-eating bird          grass

**Flow of Energy**

*I found this information on page _____.*

**Model** *an ocean food web involving the following: the Sun, penguin, krill, whale, orca, plankton, fish, seal, and squid.*

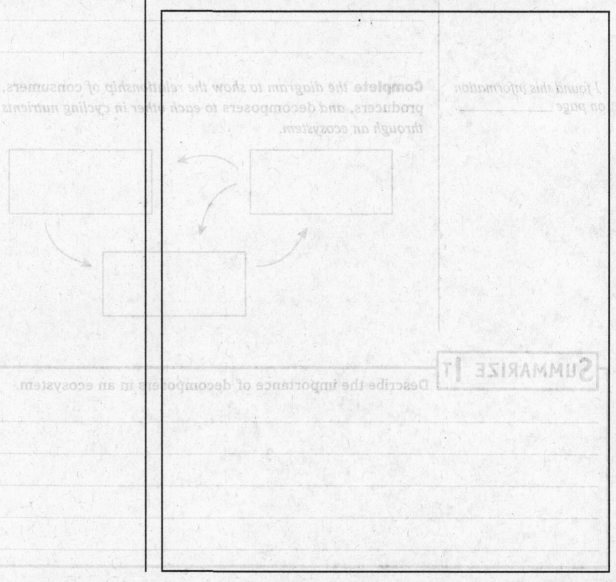

Copyright © Glencoe/McGraw-Hill, a division of The McGraw-Hill Companies, Inc.

## Section 3  Energy Through the Ecosystem (continued)

| Main Idea | Details |
|---|---|
| **Cycling of Materials**<br><br>*I found this information on page _____.* | **Summarize** *matter that can be recycled by listing three types of matter that are recycled on the lines below.*<br><br>1. _____<br><br>2. _____<br><br>3. _____<br><br>**Analyze** *why matter must be recycled through ecosystems.*<br><br>_____<br>_____<br>_____ |
| *I found this information on page _____.* | **Complete** *the diagram to show the relationship of* consumers, producers, *and* decomposers *to each other in cycling nutrients through an ecosystem.*<br><br>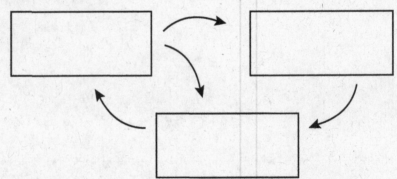 |

---

**SUMMARIZE IT**  Describe the importance of decomposers in an ecosystem.

_____
_____
_____
_____
_____

Copyright © Glencoe/McGraw-Hill, a division of The McGraw-Hill Companies, Inc.

# Tie It Together

## Make a Food Web

*With a partner, describe a habitat near where you live. Identify as many organisms as you can that share the habitat. Create a food web that shows the flow of energy through the habitat, and then change a biotic factor in the habitat. Describe the effect this change would have on the food web.*

Copyright © Glencoe/McGraw-Hill, a division of The McGraw-Hill Companies, Inc.

# Ecology   Chapter Wrap-Up

*Now that you have read the chapter, think about what you have learned and complete the table below. Compare your previous answers with these.*

1. Write an **A** if you agree with the statement.
2. Write a **D** if you disagree with the statement.

| Ecology | After You Read |
|---|---|
| • The biosphere is made up of all of the ecosystems on Earth combined. | |
| • Different species of organisms live in the same habitat. | |
| • Energy for most organisms comes from the Sun. | |
| • A producer relies on prey for its energy. | |

# Review

*Use this checklist to help you study.*

☐ Review the information you included in your Foldable.

☐ Study your *Science Notebook* on this chapter.

☐ Study the definitions of vocabulary words.

☐ Review daily homework assignments.

☐ Re-read the chapter and review the charts, graphs, and illustrations.

☐ Review the Self Check at the end of each section.

☐ Look over the Chapter Review at the end of the chapter.

**SUMMARIZE IT** After reading this chapter, identify three things that you have learned about ecology.

_____

_____

_____

_____

Copyright © Glencoe/McGraw-Hill, a division of The McGraw-Hill Companies, Inc.

# Earth's Resources

## Before You Read

*Before you read the chapter, think about what you know about the topic. List three things that you already know about Earth's resources in the first column. Then list three things that you would like to learn about them in the second column.*

| K<br>What I know | W<br>What I want to find out |
|---|---|
|  |  |
|  |  |
|  |  |

**FOLDABLES™**
**Study Organizer**

Construct the Foldable as directed at the beginning of this chapter.

### Science Journal

*Use library or online resources to learn about other uses of the Sun's energy. In your Science Journal, describe how you could use one of them.*

_____

_____

_____

_____

_____

_____

Copyright © Glencoe/McGraw-Hill, a division of The McGraw-Hill Companies, Inc.

# Earth's Resources

## Section 1 Natural Resource Use

**Scan** *the headings in Section 1 of your book. Predict three topics that will be discussed.*

1. _____

2. _____

3. _____

### Review Vocabulary

**Define** biome *using your book or a dictionary.*

biome

_____

_____

### New Vocabulary

*Use your book or a dictionary to define each vocabulary term. Then use each term in a sentence that shows its scientific meaning.*

natural resource

_____

_____

renewable resources

_____

_____

nonrenewable resources

_____

_____

### Academic Vocabulary

*Use a dictionary to define* available *to show its scientific meaning.*

available

_____

_____

Copyright © Glencoe/McGraw-Hill, a division of The McGraw-Hill Companies, Inc.

Section 1 Natural Resource Use (continued)

## Main Idea — — — Details — — Details

### News Flash: Trouble in the Rain Forest

*I found this information on page* _____

**Summarize** *why rain forests are being destroyed and what the effects of this destruction are.*

| Destruction of Rain Forests | |
|---|---|
| **Reasons for Destruction** | **Effects of Destruction** |
| To clear land for _____ _____. | By destroying habitat, species _____. |
| To harvest wood to be used as _____ | Plants that may potentially be used for medicines _____ |

### Natural Resources

*I found this information on page* _____.

**Organize** *information about natural resources in the chart below.*

| Some Natural Resources and Their Products | |
|---|---|
| **Resource** | **Products** |
| trees | |
| crude oil | |
| minerals | |
| coal | |
| plants | |

### Availability of Resources

*I found this information on page* _____.

**Identify** *the two main categories of natural resources by labeling the segments of the time line as* renewable resources *or* nonrenewable resources. *Then list at least three examples of each.*

**Years Required for Replacement by Natural Processes**

0 years                    100 years

Examples: _____        Examples: _____
_____                  _____

Copyright © Glencoe/McGraw-Hill, a division of The McGraw-Hill Companies, Inc.

## Section 1 Natural Resource Use (continued)

**Main Idea**                **Details**

*I found this information on page _____.*

**Compare** *the proportions of different resources used to meet the world's energy needs. Complete the bar graph to show the percentages contributed by five categories of resources:* crude oil, coal, natural gas, nuclear, *and* other.

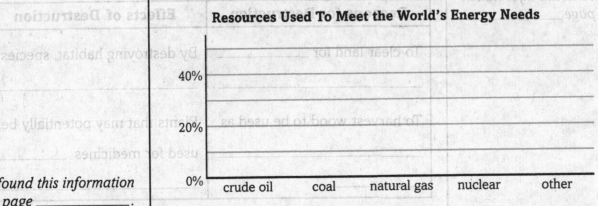

Resources Used To Meet the World's Energy Needs

40%

20%

0%

crude oil    coal    natural gas    nuclear    other

*I found this information on page _____.*

**Organize** *information about the benefits of conserving resources by completing the diagram.*

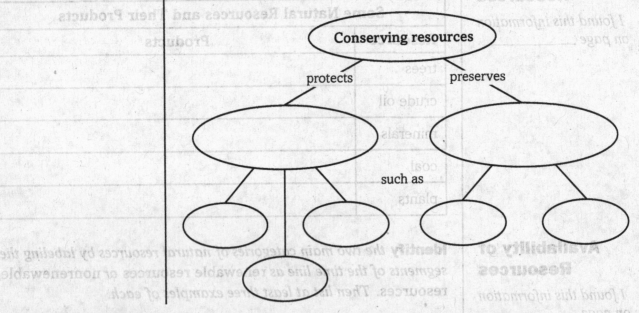

Conserving resources

protects                preserves

such as

**CONNECT IT** The bar graph that you made shows that 8 percent of the world's energy comes from resources labeled as "other." List as many resources that you can think of that are part of this category.

_____

_____

Copyright © Glencoe/McGraw-Hill, a division of The McGraw-Hill Companies, Inc.

Copyright © Glencoe/McGraw-Hill, a division of The McGraw-Hill Companies, Inc.

Name _____  Date _____

# Earth's Resources
## Section 2 People and the Environment

**Skim** *Section 2 of your book. Write three questions that come to mind. Look for answers to your questions as you read the section.*

1. _____

2. _____

3. _____

**Review Vocabulary** **Define** habitat *using your book or a dictionary.*

habitat _____

**New Vocabulary** *Read the definitions below. Write the correct vocabulary term on the blank to the left of each definition.*

_____ material that can harm living things by interfering with life processes

_____ form of pollution that occurs when gases released by burning oil and coal mix with water in the air to form rain or snow that is strongly acidic

_____ area where garbage is deposited and buried

**Academic Vocabulary** *Use a dictionary to define* eventual *to show its scientific meaning.*

eventual _____

## Section 2 People and the Environment (continued)

### ◁ Main Idea ▷

### ◁ Details ▷

**Exploring Environmental Problems and Our Impact on Land**

*I found this information on page _____.*

**Identify** *the actions that* land use laws *require before major construction takes place.*

Studies must be completed to determine the impact of construction on

---

*I found this information on page _____.*

**Describe** *two pathways that potentially* hazardous household waste *may follow upon disposal.*

Potentially hazardous material such as _____ _____ require disposal.

| | |
|---|---|
| Hazardous waste is thrown away with ordinary trash. | Hazardous waste is _____ |
| Hazardous waste may end up in a _____ where its chemicals may _____. | Hazardous waste is taken to a _____ where it is _____ |

**Our Impact on Water**

*I found this information on page _____.*

**Complete** *the statements below about sources of* water pollution.

1. _____ and _____ from farmland get into lakes and oceans.

2. Rain falling on roads and parking lots washes _____ _____ into soil and waterways.

3. Some factories sometimes release polluted water into _____.

Copyright © Glencoe/McGraw-Hill, a division of The McGraw-Hill Companies, Inc.

## Section 2  People and the Environment (continued)

### Main Idea

**Our Impact on Water**

*I found this information on page _____.*

### Details

**Summarize** *two key laws that have helped to reduce water pollution in the United States.*

| Laws to Reduce Water Pollution | |
|---|---|
| **Safe Drinking Water Act** | **Clean Water Act** |
| | |

**Our Impact on Air**

*I found this information on page _____.*

**Complete** *the flow chart below to describe how* acid rain or acid snow *forms and affects organisms.*

_____ release _____ as they are burned.

↓

The _____ mix with _____ in the air to form acid rain or snow.

The acid rain or snow falls to the ground.

The acid rain or snow falls into rivers and lakes.

↓                                      ↓

_____ can be damaged.

_____ can be harmed.

**CONNECT IT**  Some recycling centers have separate bins for disposing of fluorescent lights and televisions. Explain why this is so.

_____

_____

_____

Copyright © Glencoe/McGraw-Hill, a division of The McGraw-Hill Companies, Inc.

# Earth's Resources

## Section 3  Protecting the Environment

**Scan** *the* What You'll Learn *statements for Section 3 of your book.*
*Identify two major topics that will be discussed.*

1. _____

2. _____

### Review Vocabulary

**Define** biosphere *using your book or a dictionary.*

biosphere _____

_____

### New Vocabulary

*Use your book or a dictionary to define each vocabulary term.*
*Then use each term in a sentence that shows its scientific*
*meaning.*

solid waste _____

_____

_____

recycling _____

_____

### Academic Vocabulary

*Use a dictionary to define* item *to show its scientific meaning.*

item _____

_____

_____

Copyright © Glencoe/McGraw-Hill, a division of The McGraw-Hill Companies, Inc.

## Section 3  Protecting the Environment (continued)

**Main Idea**                    **Details**

### Cutting Down on Waste

*I found this information on page _____.*

**Compare** *the approximate amounts of different sources of solid waste produced in the United States every year by listing them in the chart below.*

| Sources of Solid Waste | |
|---|---|
| **Source** | **Approximate Amount (millions of tons)** |
| Aluminum | |
| Other metals | |
| Glass | |
| Plastics | |
| Yard waste | |
| Paper products | |
| Other waste | |

*I found this information on page _____.*

**Define** *what is meant by* reduce *in* reduce, reuse, recycle.

"Reduce" refers to _____

_____. Compared to other ways to

help solve the solid waste problem, this is _____

_____.

*I found this information on page _____.*

**Complete** *the graphic organizer by providing at least three ways to reduce the total amount of solid waste that you throw away.*

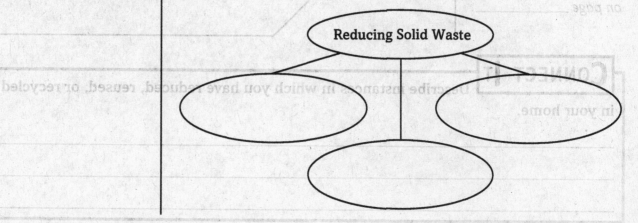

Copyright © Glencoe/McGraw-Hill, a division of The McGraw-Hill Companies, Inc.

## Section 3 Protecting the Environment (continued)

⟨ **Main Idea** ⟩    ⟨ **Details** ⟩

*I found this information on page _____.*

**Create** *a list of at least eight items that can be* reused.

| Examples of Reusable Items |
|----------------------------|
|                            |
|                            |
|                            |

*I found this information on page _____.*

**Connect** *products to new products into which they can be* recycled.

**Recycling Products**

Original Products          Recycled Products

| plastic bottles |  |  |
|---|---|---|

| glass |  |  |
|---|---|---|

| leftover food |  |  |
|---|---|---|

### Habits for a Healthier Environment

*I found this information on page _____.*

**Identify** *the benefits of practicing the* Three Rs.

Practicing the Three Rs helps to

┌─────────────────────────────────────────────────┐
│ **CONNECT IT** │
│ Describe instances in which you have reduced, reused, or recycled │
│ **in your home.** │
│ _____ │
│ _____ │
│ _____ │
└─────────────────────────────────────────────────┘

Copyright © Glencoe/McGraw-Hill, a division of The McGraw-Hill Companies, Inc.

# Tie It Together

## Conserving at School

*Think about all of the things that you do in an ordinary school day. Write these activities down on the left half of a piece of paper beginning with getting up in the morning to the sound of an alarm clock. Then, on the right side of the list, write what you could do to accomplish each of these activities using less energy or fewer resources. Make sure that you note where you can reuse, recycle, or reduce the amount of materials you use.*

_____  _____
_____  _____
_____  _____
_____  _____
_____  _____

_____  _____
_____  _____
_____  _____
_____  _____
_____  _____
_____  _____
_____  _____

_____  _____
_____  _____
_____  _____
_____  _____

Copyright © Glencoe/McGraw-Hill, a division of The McGraw-Hill Companies, Inc.

# Earth's Resources Chapter Wrap-Up

*Review the ideas you listed in the chart at the beginning of the chapter. Cross out any incorrect information in the first column. Then complete the chart by filling in the third column.*

| K What I know | W What I want to find out | L What I learned |
|---|---|---|
|  |  |  |

# Review

*Use this checklist to help you study.*

☐ Review the information you included in your Foldable.

☐ Study your *Science Notebook* on this chapter.

☐ Study the definitions of vocabulary words.

☐ Review daily homework assignments.

☐ Re-read the chapter and review the charts, graphs, and illustrations.

☐ Review the Self Check at the end of each section.

☐ Look over the Chapter Review at the end of the chapter.

## SUMMARIZE IT
After reading this chapter, identify three main ideas from the chapter.

_____

_____

_____

Copyright © Glencoe/McGraw-Hill, a division of The McGraw-Hill Companies, Inc.

**accumulate:** to increase gradually in quantity or number; to gather or pile up

**affect:** to bring about a change in

**apparent:** appearing as actual

**available:** suitable or ready for use or service or at hand; readily obtainable or accessible

**category:** group or class of things

**chart:** organizational tool that gives information about something in the form of a diagram, graph, or table

**chemical:** acting to change the identity, or chemical makeup, of a substance

**complex:** made up of complicated and related parts

**compound:** substance produced when elements combine and whose properties are different from each of the elements in it

**constant:** continual; going on all the time

**contact:** act or state of touching or meeting

**contract:** to become smaller

**convert:** to change from one form or use to another

**cycle:** series of events or actions that repeat regularly

**decline:** to become less in health, power, value, or number

**design:** to build or create to satisfy a need

**distribute:** to divide among several or many things; scatter

**encounter:** to meet or experience

**erode:** to wear away

**evaluate:** to carefully judge the significance of something

**eventual:** taking place at an unspecified later time

**exert:** to bring to bear

**expose:** to reveal or make known

**factor:** something that contributes to a result

**feature:** part, appearance, or characteristic of something

**framework:** supporting structure

**function:** special work or purpose of an object or a person

**goal:** objective or end that one strives to achieve

**identical:** exactly the same

**identify:** to recognize or show to be a person or thing that is known

**indicate:** to point out, give evidence of, or show

**inject:** to force into something

**injure:** to cause bodily harm

**input:** power or energy that is put into a machine or system

**interact:** to act on each other

**internal:** happening or arising or located within

Copyright © Glencoe/McGraw-Hill, a division of The McGraw-Hill Companies, Inc.

# Academic Vocabulary

**item:** distinct part in an enumeration, account, or series; an object of attention, concern, or interest

**layer:** one thickness over another

**maintain:** to keep or preserve in an existing state

**medium:** substance through which a force or effect is transmitted

**neutral:** neither negative nor positive

**occur:** to take place or happen

**overlap:** one thing extends over another

**parallel:** being the same distance apart at all points

**phase:** stage of development

**process:** series of natural changes that cause a particular result

**range:** the difference between the highest and lowest values

**react:** to undergo a chemical change

**refine:** to separate from impurities

**rigid:** stiff, inflexible

**segment:** piece or separate part of something; to separate into parts or sections

**significant:** important

**similar:** having many but not all qualities alike

**summary:** brief account that covers the main points

**symbol:** something that represents something else

**technology:** use of science for practical purposes, especially in engineering and industry

**temporary:** not permanent or lasting

**theory:** explanation of things or events based on scientific knowledge resulting from many observations and experiments

**transfer:** to carry or send from one person, place, or position to another

**transform:** to change the condition, nature, or function of; to convert

**unique:** one of a kind

**visible:** capable of being seen

Copyright © Glencoe/McGraw-Hill, a division of The McGraw-Hill Companies, Inc.